W9-AXX-131

COMPUTER SIMULATION OF BIOMOLECULAR SYSTEMS

THEORETICAL AND EXPERIMENTAL APPLICATIONS

COMPUTER SIMULATION OF BIOMOLECULAR SYSTEMS

THEORETICAL AND EXPERIMENTAL APPLICATIONS

Proceedings of two colloquia organized by Alliant Computer Systems Corporation, December 16-18, 1987, Princeton, New Jersey, U.S.A. and April 20-23, 1988, Amsterdam, The Netherlands

Edited by

Wilfred F. van Gunsteren
Department of Physical Chemistry, University of Groningen,
9747 AG Groningen, The Netherlands and
Department of Physics, Free University, 1007 MC Amsterdam,
The Netherlands

and

Paul K. Weiner
Alliant Computer Systems Corporation,
Littleton, Massachusetts 01460, U.S.A.

ESCOM Leiden 1989

CIP-Data Koninklijke Bibliotheek, Den Haag

Computer

Computer Simulation of Biomolecular Systems: Theoretical and Experimental Applications / [Eds.] W.F. van Gunsteren, P.K. Weiner. - Leiden : ESCOM. - Ill.
The volume consists of introductory papers, research articles and reviews from invited speakers who participated in two scientific colloquia organized by Alliant Computer Systems Corporation on December 16-18, 1987 in Princeton (USA) and April 20-23, 1988 in Amsterdam (The Netherlands). - With ref.
SISO 527.8 UDC 681.3:577
Subject headings: Computer simulation / Biomolecular systems.

ISBN 90-72199-03-0 (hardbound)

Published by:

ESCOM Science Publishers B.V.
P.O. Box 214
2300 AE Leiden
The Netherlands

Printed in The Netherlands

Preface

During the academic year 1987/1988, Alliant Computer Systems Corporation sponsored and organized two scientific colloquia on computational methods in biomolecular chemistry. The first one, held in Princeton, New Jersey (USA), December 16-18, 1987, was titled 'Free Energy Perturbation' and comprised contributions from Beveridge, Lybrand, Singh, Jorgensen, Bash, Hagler, Karplus, Brooks, Pettitt, Goddard, Kollman, Van Gunsteren and Hermans. It drew about 150 participants and constituted a very timely state-of-the-art overview of the successes and also of the problems concerning methods to compute free energies in biomolecular systems by computer simulation. The second colloquium, held in Amsterdam (The Netherlands), April 20-23, 1988, was titled 'Computer Simulations in Protein Engineering and Drug Design' and comprised contributions from Beveridge, Gerber, Pettitt, Kollman, Kaptein, Dobson, Warshel, Goddard, Maggiora, Tilton, Hol, Karplus, Lehn, Wells, Holtzman, Blundell and Van Gunsteren. It also drew about 150 participants. This time the subject was broader and covered both theoretical and experimental aspects of protein engineering and drug design.

The considerable interest in these meetings led Alliant Computer Systems to consider publication of the proceedings of the colloquia in such a manner that it would form:

(1) A concise introduction to the technique of free energy computation;
(2) An account of the state-of-the-art as far as the application of these techniques to molecular systems is concerned; and
(3) An account of a few other recently developed computational techniques in the field of molecular modeling and structure refinement.

The speakers of both colloquia were invited to submit a contribution to this book. Those who responded had their papers reviewed by the editors. The best of the submitted papers are included in this volume. Since the number of papers are, because of the selection process, limited, the book will not constitute a comprehensive overview of the subjects mentioned in the titles of the meeting. It does include a sampling of introductory papers, research articles and review papers that are of relevance to both experimental and theoretical researchers interested in these topics. Its character is much like that of the proceedings of two earlier meetings on the various aspects of computer simulation of chemical and biomolecular systems [1,2].

Beveridge and DiCapua give a clear introduction to the statistical-mechanical basis of methods to calculate free energy. Van Gunsteren provides an overview of approximations, practical considerations and limitations concerning this type of calculation. Jorgensen presents an instructive review of the contributions of his group in the area of free energy computation. The paper of Brooks contains illustrative examples of the calculation of free energies of hydration and of protein-inhibitor binding. Lybrand subsequently and very briefly discusses free energy aspects of drug-DNA binding. In the next contribution, Pettitt critically considers the accuracy of calculated free energies. This is also the subject of the contribution of Pearlman and Kollman. Warshel and Creighton provide a description of the capabilities of their computer programs and a list of references to Warshel's work on computer simulation. The series of papers on various aspects of free energy calculations is completed by some thoughts of Jan Hermans on the past and future of free energy simulations of biological macromolecules.

Hagler and coworkers discuss a number of aspects of potential energy functions for organic and biomolecular systems. The next two contributions concern molecular modeling. Hubbard and Blundell give an account of designing a novel protein. Gerber deals with the generation of conformations for cyclic structures. The final two contributions are from experimentally oriented groups. Gros and coworkers demonstrate the power of crystallographic refinement of a protein based on homology and molecular dynamics simulation techniques. Kaptein and his group give an overview of the technique by which spatial structure of biomolecules can be obtained from NMR.

This series of papers illustrates how much has been achieved during the last few years in the area of computer simulation of biomolecular systems, but also, most importantly, that much is still to be investigated and learned. They form a, though necessarily incomplete, state-of-the-art perspective on the capabilities and limitations of computer simulation techniques for the study of biomolecular systems.

We would like to thank Alliant Computer Systems for organizing both colloquia, the participants for providing stimulating meetings, and the contributors for their papers.

Wilfred F. van Gunsteren
Paul K. Weiner

References

1. Hermans, J. (Ed.) Molecular Dynamics and Protein Structure, Polycrystal Book Service, P.O. Box 27, Western Springs, IL 60558, 1985.
2. Beveridge, D.L. and Jorgensen, W.L. (Eds.) Computer Simulation of Chemical and Biomolecular Systems, Ann. N.Y. Acad. Sci., 482 (1986).

Contents

Preface v

Free energy via molecular simulation: A primer 1
D.L. Beveridge and F.M. DiCapua

Methods for calculation of free energies and binding constants: Successes and problems 27
W.F. van Gunsteren

Free energies in solution: The aqua vitae of computer simulations 60
W.L. Jorgensen

Thermodynamic calculations on biological systems 73
C.L. Brooks III

Free energy perturbation calculations in drug design applications 89
T.P. Lybrand

Successes, failures and curiosities in free energy calculations 94
B.M. Pettitt

Free energy perturbation calculations: Problems and pitfalls along the gilded road 101
D.A. Pearlman and P.A. Kollman

Microscopic free energy calculations in solvated macromolecules as a primary structure-function correlator and the MOLARIS program 120
A. Warshel and S. Creighton

Thoughts about the past and future of free energy simulations of biological macromolecules 139
J. Hermans

Potential energy functions for organic and biomolecular systems 149
A.T. Hagler, J.R. Maple, T.S. Thacher, G.B. Fitzgerald and U. Dinur

The design of novel proteins using a knowledge-based approach to computer-aided modeling 168
T.J.P. Hubbard and T.L. Blundell

Shape-guided generation of conformations for cyclic structures 183
P.R. Gerber

Molecular dynamics refinement of the X-ray structure of thermitase complexed with eglin-c 190
P. Gros, M. Fujinaga, B.W. Dijkstra, K.H. Kalk and W.G.J. Hol

Biomolecular structures from NMR: Computational aspects 194
R. Kaptein, R. Boelens and J.A.C. Rullmann

Subject index 217

Free energy via molecular simulation: A primer

D.L. Beveridge and F.M. DiCapua
Departments of Chemistry and Molecular Biology and Biochemistry, Wesleyan University, Middletown, CT 06457, U.S.A.

1. Introduction

Molecular simulation can be defined as the numerical determination of the statistical thermodynamics and related structural, energetic and dynamical properties of a molecular assembly from an ensemble of diverse geometrical configurations of the system generated on a high-speed digital computer. Computer simulations on molecular systems can be carried out in a probabilistic mode using Monte Carlo (MC) methods first described in 1953 by Metropolis et al. [1], or deterministically following the Newtonian methods initially set forth in the early 1960s by Alder and Wainwright [2] and Rahman and Stillinger [3], known as molecular dynamics (MD). Applications of molecular simulation to chemical and biomolecular systems currently range from the investigation of the structure and energetics of condensed fluid phases such as liquid water and aqueous solutions to diverse molecular liquids and nonaqueous solutions [4], and from initial studies of protein dynamics to the study of molecular motions in DNA as well as the dynamics of drug-DNA interaction [5]. Preliminary studies of more complex systems such as protein-DNA interactions and even biological membrane structure have been reported, and numerous biological systems are expected to be accessible to simulation studies in the near future.

The quantities determined in a typical Monte Carlo or molecular dynamics simulation include the average or mean configurational energy of the system (formally identified with the thermodynamic excess internal energy), the various spatial distribution functions for equilibrium systems and time correlation functions for dynamical systems, and the diverse indices of a structural and energetical analysis of these functions. Although conventional forms of molecular simulations have been very useful, diverse problems in structural and reaction chemistry of molecules in solution such as solvation potentials, solvent effects on conformational stability, chemical reaction mechanisms and dynamics as studied via absolute reaction rate theory and ligand binding to macromolecules require, in addition to the properties and related quantities routinely available, a particular knowledge of the configurational free energy of the system. This quantity in principle follows directly from the statistical thermodynamic partition function for the system. In practice, the determination of free energy with sufficient numerical accuracy turns out to be problematic in molecular simulation, and hence the need for a special treatment of this subject.

To appreciate the problem, it must be recognized that considerations on free energy in molecular simulations take a distinctly different form for intramolecular and intermolecular degrees of freedom. For the intramolecular case, the problem involves vibrational and librational modes of motion on the intramolecular energy surface. In the vicinity of well-defined local energy minima, the free energy is readily accessible and can be computed by constructing a partition function from the vibrational frequencies obtained from a normal mode calculation and treating the problem in the harmonic approximation, essentially an extension of the Einstein oscillator formalism. This has been successfully applied to polypeptide systems by Hagler et al. [6]. Anharmonic effects can be introduced by performing a molecular simulation on the intramolecular energy surface using Monte Carlo or molecular dynamics, and calculating the entropy in the quasi-harmonic approximation as described by Karplus and Kushick [7].

In fluids, including molecular liquids, solutions and vapors, the particles of the system undergo diffusional motion, and a harmonic or quasiharmonic approximation breaks down. These considerations apply also in the case of a flexible molecule, in which conformational transitions are effectively an intramolecular 'diffusional mode'. Conventional Monte Carlo and molecular dynamics procedures for diffusional modes, although firmly grounded in Boltzmann statistics, do not ordinarily proceed via the direct determination of a partition function. The exponential dependence of the Boltzmann factor on the energy makes this quantity notoriously slow to converge in N-dimensional space for densities typical of molecular and biomolecular systems. The Monte Carlo-Metropolis method is in fact a Markov process designed specifically to avoid the necessity of determining a partition function in the calculation of internal energy and related thermodynamic properties in a simulation. In molecular dynamics, the physical nature of the calculated trajectories of the particles serves this purpose equally as well, with a validity assured by the essential ergodicity of the system. In any case, in the absence of a partition function, one is unable to compute the free energy of the system straightforwardly when diffusional modes are involved.

Recently, special methodologies have been introduced to provide alternative routes to free energy calculations in molecular simulations. However, computationally more intensive numerical procedures are required in all but the simplest of cases. With the advent of supercomputers, dedicated purpose minisupercomputers and multiprocessors, these computational barriers are being overcome, and free energy determinations in problems on chemical and biomolecular systems of considerable interest and importance have become quite feasible. Diverse novel and interesting results are currently being obtained, many of which are described in the succeeding articles in this volume.

The purpose of this paper is to provide an elementary introduction to free energy determination via molecular simulation. A basic familiarity with statistical thermodynamics and molecular simulation theory is presumed, but we have tried to make this article as self-explanatory as possible. For further theoretical background, the interested reader is referred to the new comprehensive text on molecular simulation by Allen and Tildesley [4] and the monograph on molecular dynamics of proteins and nucleic acids by McCammon and Harvey [5]. The interested reader is also referred to the other recent reviews related to this subject by Valleau and Torrie [8], Quirke [9], Shing and Gubbins [10], Pohorille and Pratt [11] and by Mezei and Beveridge [12]. Deeper background on free energy methods in liquid state theory is given in an earlier comprehensive review by Barker and Henderson [13].

The following section collects the relevant background material on the subject, particularly on the statistical thermodynamic quantities involved, and on relevant molecular simulation protocols. The free energy problem is then discussed in more mathematical terms, and the coupling parameter approach, a useful mathematical construct, is introduced. Each of the main methods for determining free energy numerically in molecular simulation can be expressed as a simple manipulation of the free energy of the system expressed as a function of coupling parameter, and are the subject of Sections 3-5. In each section, a general formulation of the method is given, followed by a discussion of methodology. Then citations to several instructive and exemplary research papers from the earlier literature on this subject are given, each of which is well worth careful study, in parallel with this article, by those interested in becoming practitioners of this art-form. The serious reader should have copies of these articles readily at hand. Our intent here is by no means to review the literature; leading references to studies on chemical and biomolecular systems through 1986 are compiled in Ref. 12. Recent and current applications to biomolecular systems will be reviewed by the authors in the forthcoming (Vol. 88) edition of the *Annual Review of Biophysics and Biophysical Chemistry.*

2. Background

(A) Statistical thermodynamic quantities

Important statistical thermodynamic quantities for a discussion of computer simulations and the free energy problem, referred to the canonical ensemble, are

(a) The configurational partition function, Z, of the system

$$Z_N = \int \cdots \int \exp[-\beta E(\mathbf{X}^N)] d\mathbf{X}^N \tag{1}$$

where $E(\mathbf{X}^N)$ is the configurational energy, $\beta = (kT)^{-1}$, k is Boltzmann's constant, T is the absolute temperature, and the configurational integration extends over all space $d\mathbf{X}^N$ of the N-particle system;

(b) The Boltzmann probability functional

$$P(\mathbf{X}^N) = \frac{\exp[-\beta E(\mathbf{X}^N)]}{Z_N} \tag{2}$$

for a configuration corresponding to a particular energy, $E(\mathbf{X}^N)$;

(c) The average or mean energy expression

$$U = \int \cdots \int E(\mathbf{X}^N)\, P(\mathbf{X}^N)\, d\mathbf{X}^N = \langle E(\mathbf{X}^N) \rangle \tag{3}$$

equivalent to the excess internal of the system (note the bracket notation for an ensemble average);

(d) The heat capacity

$$C_V = \frac{\langle E(\mathbf{X}^N)^2 \rangle - \langle E(\mathbf{X}^N) \rangle^2}{kT^2} \tag{4}$$

and

(e) The excess or configurational free energy

$$A = -kT \ln \frac{Z}{(8\pi^2 V)^N} \tag{5}$$

which is to be the essential focus of interest herein. The appearance of the normalization factor $(8\pi^2V)^N$ in the denominator of Eq. 5 establishes that our quantity A, a Helmholtz free energy, is defined relative to an ideal gas reference state, $E(\mathbf{X}^N) = 0$. For an atomic fluid, or a macromolecule treated as interacting atoms, the orientational factor $8\pi^2$ is replaced by unity. The corresponding Gibbs free energy is, of course,

$$G = A + PV = U + PV - TS \tag{6}$$

where P is the pressure of the system and U + PV is the enthalpy, H. All quantities and methodology described here can be readily formulated in the constant pressure (T, P, N) ensemble in which G is the thermodynamic potential as well as the canonical (T, V, N) ensemble. The grand canonical ensemble invites alternative interesting approaches to the problem (for leading references, see Ref. 12), which have not yet proven to be of practical utility for chemical and biomolecular systems.

(B) Protocols of molecular simulation

A computer simulation on a molecular system, Monte Carlo or molecular dynamics, begins with an arbitrarily chosen initial configuration, and involves generating an ensemble of structures by well defined prescriptions. In the initial segment of the simulation, the energy of the system (potential in MC, kinetic energy in MD*) changes rapidly, but eventually settles into a reasonable approximation of an oscillation about a mean, at which point the simulation has achieved 'equilibration'. For a fully ergodic system, in which all regions of phase space are accessible, the region of equilibration will be the same regardless of the choice of initial configuration. At the point where this has been numerically achieved, the previous history of the calculation is discarded, and the 'production' phase commences. Calculated properties are formed from the averages over the ensemble of structures generated during the production phase of a simulation, usually millions of configurations in a Monte Carlo calculation and tens of picoseconds in molecular dynamics on present generation machines. The simulation is allowed to proceed until the calculated properties are no longer significantly changing, at which point the simulation is said to be stabilized, or 'converged'.

Configuration space is by definition of order infinity, and since a given computer simulation is by definition finite, one obtains from a simulation of finite length only an estimate of the true, infinite-order ensemble averages of the energy and other quantities determined in the simulation. Another finite segment of the realization, commencing with a different but perfectly acceptable equilibrated structure, would be expected to lead to a slightly different estimate. The set of possible estimates from various realizations in a proper simulation are independent random estimates of the true mean, and should be normally

*In practice, free energy determinations via molecular dynamics are sometimes carried out in an isothermal ensemble with equilibration not well defined. We maintain here the microcanonical representation in order to parallel the Monte Carlo method more closely.

distributed. In an important early article on Monte Carlo studies of simple liquids, Wood [14] applied standard small sample statistical theory to the problem. A realization consisting of n total steps is broken up into some small number m of successive sequences (blocks, batches) of p steps each to give

$$U_i = \frac{1}{p}\sum_{k=1}^{p} E(X_k^N) \qquad X_k \in \text{MC or MD}, \qquad i=1, m \tag{7}$$

where the U_i are the block averages.

In terms of the m block averages, the total internal energy is

$$U = \frac{1}{m}\sum_{i=1}^{m} U_i \tag{8}$$

The statistical uncertainty in the computed internal energy U is then simply estimated from the standard deviation

$$\Delta U = \left[\sum_{i=1}^{m} \frac{(U_i - U)^2}{m(m-1)}\right]^{1/2} \tag{9}$$

and 67% of an infinite number of estimates would fall within $\pm\Delta U$ of U. The size of the blocks used in this calculation is in principle determined so as to satisfy certain criteria designed to assure the U's are asymptotically mutually independent of one another. The block size is large enough when ΔU is observed to be unchanged upon further increase of p. A display of both cumulative mean energy and the batch means for segments along the realization as a function of configurations in the sampling gives a simple statistical tool for assessing the stability, if not strictly speaking the convergence, of the calculation. Consideration of the statistical uncertainties as computed by Eq. 9 is vital in the analysis and interpretation of the results of a simulation, and in a critical comparison of calculated and experimentally observed values. Issues of convergence in various other ensemble averages also figure significantly in the free energy problem, and crop up frequently in the following discussion.

Finally, it should be noted that each property of the system will have its own characteristic convergence profile. Simple average quantities, such as mean energy, are the simplest to compute and the fastest to converge. Fluctuation properties, such as heat capacity, are well-known to achieve convergence slower than mean energy. The convergence of each individual property can be monitored in the course of the simulation, and the method of batch means described above can be used to assign appropriate statistical uncertainties to each calculated quantity. Further discussion of the statistical uncertainty problem in molecular simulations has been given by Straatsma et al. [15].

(C) The free energy problem

With the preceding exposition of background, it becomes possible to specify mathematically the problem encountered in free energy determination using molecular simulation. First, let us cast the free energy into the form of an ensemble average expression adaptable to evaluation via simulation in a manner analogous to the internal energy. Inserting unity in the form of

$$1 = (8\pi^2 V)^{-N} \int \ldots \int \exp[+\beta E(\mathbf{X}^N)] \exp[-\beta E(\mathbf{X}^N)]\, d\mathbf{X}^N \tag{10}$$

into the argument of the logarithm in Eq. 5, and inverting, results in

$$A = kT \ln \left[\frac{\int \ldots \int \exp[+\beta E(\mathbf{X}^N)] \exp[-\beta E(\mathbf{X}^N)]\, d\mathbf{X}^N}{\int \ldots \int \exp[-\beta E(\mathbf{X}^N)]\, d\mathbf{X}^N} \right] \tag{11}$$

$$A = kT \ln \int \ldots \int \exp[+\beta E(\mathbf{X}^N)]\, P(\mathbf{X}^N)\, d\mathbf{X} \tag{12}$$

or, using the bracket notation,

$$A = kT \ln < \exp[+\beta E(\mathbf{X}^N)] > \tag{13}$$

In principle, this equation provides a means for the calculation of excess free energy in one conventional simulation determining an ensemble of configurations consistent with the probability function, $P(\mathbf{X}^N)$, and integrating over all space as in the determination of average energy via Eq. 3. However, expanding the exponential, one obtains

$$\exp[+\beta E(\mathbf{X}^N)] = 1 + \beta E(\mathbf{X}^N) + \frac{1}{2} \beta^2 E(\mathbf{X}^N)^2 + \ldots \tag{14}$$

and the ensemble energy average involved in the free energy equation becomes

$$< \exp[+\beta E(\mathbf{X}^N)] > = 1 + \beta < E(\mathbf{X}^N) > + \ldots \tag{15}$$

The logarithm of the ensemble average expression expands as

$$\ln[1+\beta<E(\mathbf{X}^N)>+...] = \beta[<E(\mathbf{X}^N)> - \frac{\beta}{2!}<E(\mathbf{X}^N)>^2+...] \quad (16)$$

and the corresponding free energy expansion is thus

$$A = <E(\mathbf{X}^N)> - \frac{\beta}{2!}<E(\mathbf{X}^N)>^2+... \quad (17)$$

where the first term is internal energy, and the remainder of the expansion is equivalent to –TS, where S is the entropy of the system.

This indicates clearly that the evaluation of ensemble averages of the exponential is equivalent to determining the ensemble averages of not only the configurational energy, but also the energy squared, as encountered in heat capacity determinations, and other higher powers of the energy (fluctuation terms) to an extent depending upon the radius of convergence of the expansion. Thus, the free energy computed in a Monte Carlo-Metropolis or a molecular dynamics simulation by the ensemble average, Eq. 14 is bound to be slower to converge than simply the average energy. Since convergence of the average energy is already time consuming to achieve, free energy will be somewhat worse in this regard, and therein lies the difficulty.

The free energy problem can be appreciated from another perspective by examining the configurational integral explicitly,

$$A = kT \ln\int...\int \exp[+\beta E(\mathbf{X}^N)]\, P(\mathbf{X}^N)\, d\mathbf{X}^N \quad (18)$$

where $P(\mathbf{X}^N)$ is proportional to $\exp[-\beta E(\mathbf{X}^N)]$, as given in Eq. 2. Because of the rapid increase of $\exp[+\beta E(\mathbf{X}^N)]$ with energy, higher energy regions of configuration space are expected to be important to the configurational integral of Eq. 14. However, molecular simulation by either Monte Carlo-Metropolis or molecular dynamics, sampling on a probability measure proportional to $\exp[-\beta E(\mathbf{X}^N)]$, i.e. $P(\mathbf{X}^N)$, seeks out preferentially the *lower* regions of configuration space. For densities typical of molecular liquids, the high- and low-energy regions of configurational space are sufficiently well separated that a Monte Carlo or molecular dynamics realization of practical length will never adequately sample the high-energy regions that are expected to contribute most to the ensemble average required. Thus, calculating the absolute free energy via Eq. 18 in conventional molecular simulation is likely to lead to poorly converged, and thus inaccurate, numerical estimates of this quantity.

(D) The coupling parameter approach

In molecular systems we are frequently confronted not only with the determination of the absolute excess free energy, A, but the free energy difference,

ΔA, between two well-defined states. Here, for states denoted 0 and 1, ΔA is given in terms of the ratio of the partition functions for the two states, Z_1 and Z_0 as

$$\Delta A = A_1 - A_0 = -kT \ln \frac{Z_1}{Z_0} \tag{19}$$

A straightforward approach to the free energy difference ΔA, would require independent determinations of Z_0 and Z_1 based on energy functions E_0 and E_1 (the dependence on configurational coordinates is implicit) which individually are subject to all the numerical difficulties detailed in the preceding section. A number of theoretical and methodological approaches to the problem can be cast in the form of a very useful construct, the 'coupling parameter'. Let us assume that the potential, E, depends on a continuous parameter λ, such that as λ is varied from 0 to 1, $E(\lambda)$ passes smoothly from E_0 to E_1. This defines an analytical continuation

$$A(\lambda) = -kT \ln Z(\lambda) \tag{20}$$

The calculation of ΔA can be performed by integrating the derivative of $A(\lambda)$ along λ (thermodynamic integration), by designating intermediate states $A(\lambda_i)$ spaced closely enough and use a form of Eq. 13 to compute ΔA in a stepwise manner (perturbation method) or actually developing $A(\lambda)$ on the [0,1] interval from simulations where λ is variable, and obtaining ΔA as $A(\lambda=1)-A(\lambda=0)$. The function, $A(\lambda)$, is equivalent to the 'potential of mean force' with respect to the coordinate λ in the statistical thermodynamics of fluids. Thermodynamic integration, the perturbation method, and the potential of mean force calculations are the main approaches to free energy determinations via molecular simulation, and are each in turn the subject of the succeeding sections of this article.

In a modern sense, the coupling parameter approach originates in the derivation of an important integral equation in liquid state theory by Kirkwood [16], but the seeds of this idea can be traced back to the work of De Donder on chemical affinity, and the degree of advancement parameter for a chemical process [17]. The coupling parameter, λ, is a generalized extent parameter, and defines a path between intial and final states 0 and 1. On a physical path, a knowledge of $A(\lambda)$ can be used to determine the free energy of activation $\Delta A^{\ddagger}$, as well as ΔA. It is important to note that nonphysical paths are admissible in the case where the quantity of interest is a state function like ΔA, which of course is independent of the path. This turns out to be a critical advantage in doing actual problems, since a nonphysical pathway may be computationally convenient. An important example of this comes in the case of a 'mutational' process where relative free energies for a series of molecules can be computed by using a λ-coordinate to computationally change one functional group, subunit or even substrate into another. Overall, one must be aware that there is considerable freedom in the choice of the λ-coordinate, and decisions on the selection of λ are usually made by combining physical and numerical requirements.

The changes involved in going from an initial to a final state of a chemical system involve, in many cases of interest, changes in the molecular topography. A categorization of the various types of changes on an λ-coordinate in terms of 'topographical transition coordinates' has been set forth in Ref. 12, and is summarized schematically in Fig. 1.

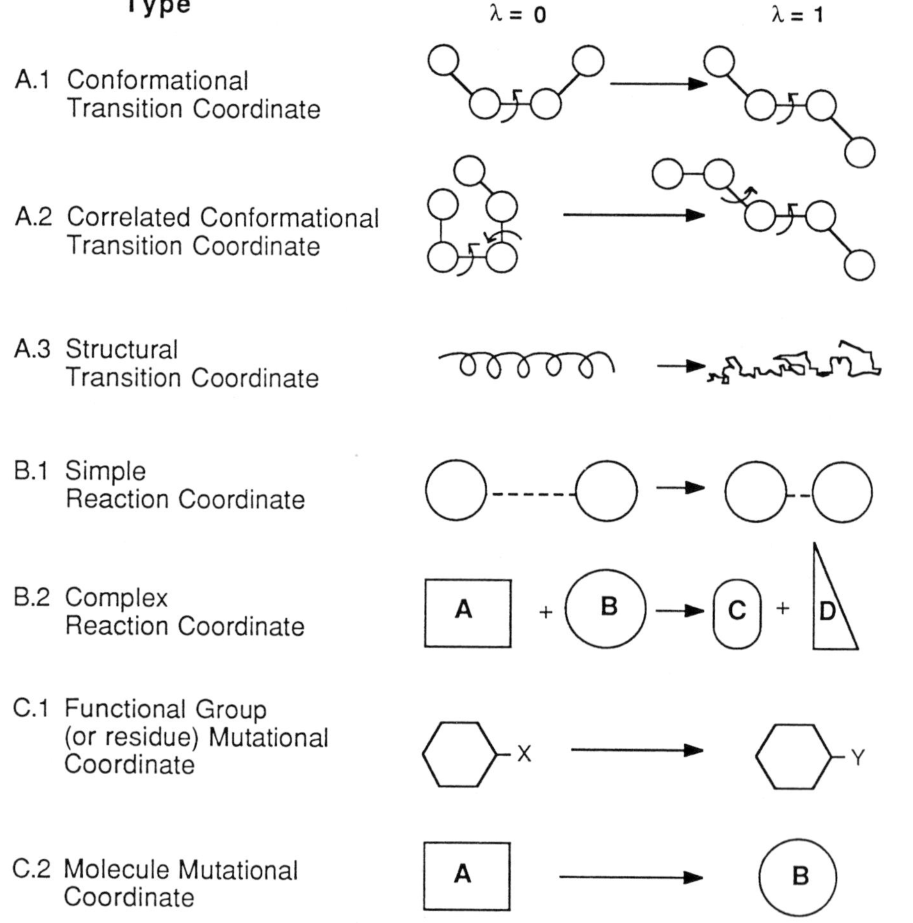

Fig. 1. Types of topographical transition coordinates used in λ-integrations for free energy determinations via molecular simulation.

3. Thermodynamic integration

(A) General formulation

Given the free energy function, A(λ), for a system defined on an interval of λ from an initial state λ=0 to a final state λ=1, the corresponding free energy difference between these states follows simply from the definition of an integral in elementary calculus as

$$\Delta A = \int_0^1 \frac{\partial A(\lambda)}{\partial \lambda} \, d\lambda \tag{21}$$

This integral can be conveniently expressed as an ensemble average as follows. If

$$A(\lambda) = -kT \ln Z(\lambda) \tag{22}$$

then

$$\frac{\partial A(\lambda)}{\partial \lambda} = -kT \left[\frac{\partial \ln Z(\lambda)}{\partial \lambda} \right] = \frac{-kT}{Z(\lambda)} \frac{\partial Z(\lambda)}{\partial \lambda} \tag{23}$$

Now

$$Z(\lambda) = \int \ldots \int \exp[-\beta E(\mathbf{X}^N, \lambda)] \, d\mathbf{X}^N \tag{24}$$

and

$$\frac{\partial Z(\lambda)}{\partial \lambda} = \int \ldots \int \frac{\partial}{\partial \lambda} \exp[-\beta E(\mathbf{X}^N, \lambda)] \, d\mathbf{X}^N \tag{25}$$

$$\frac{\partial Z(\lambda)}{\partial \lambda} = -\beta \int \ldots \int \frac{\partial E(\mathbf{X}^N, \lambda)}{\partial \lambda} \exp[-\beta E(\mathbf{X}^N, \lambda)] \, d\mathbf{X}^N \tag{26}$$

Thus, substituting back into Eq. 23,

$$\frac{\partial A(\lambda)}{\partial \lambda} = \frac{1}{Z(\lambda)} \int \ldots \int \frac{\partial E(\mathbf{X}^N, \lambda)}{\partial \lambda} \exp[-\beta E(\mathbf{X}^N, \lambda)] \, d\mathbf{X}^N \tag{27}$$

or

$$\frac{\partial A(\lambda)}{\partial \lambda} = \left\langle \frac{\partial E(\mathbf{X}^N,\lambda)}{\partial \lambda} \right\rangle_\lambda \tag{28}$$

where the subscript, $_\lambda$, indicates an ensemble average over the 'λ probability function'.

$$P(\mathbf{X}^N,\lambda) = \frac{\exp[-\beta E(\mathbf{X}^N,\lambda)]}{\int \ldots \int \exp[-\beta E(\mathbf{X}^N,\lambda)]\, d\mathbf{X}^N} \tag{29}$$

that is, the probability of finding the system in configuration $\mathbf{X}^N$ at a position λ between the initial and final state of the transformation. Finally,

$$\Delta A = \int_0^1 \left\langle \frac{\partial E(\mathbf{X}^N,\lambda)}{\partial \lambda} \right\rangle_\lambda d\lambda \tag{30}$$

The direct numerical evaluation of Eq. 30 in molecular simulation, integrating between initial and final states using some thermodynamic relationship, is called 'thermodynamic integration'. Textbook cases in physical chemistry are recovered when λ is identified with a volume change or a temperature change in the system, giving rise to an integral involving the virial expression for pressure and the Gibbs-Helmholtz equation, respectively. A point particularly important to the development herein is that the integration variable, λ, need not be restricted to simply thermodynamic variables like volume and temperature, but can be used to define an analytical continuation between initial and final states in many conceivable ways. This idea finds broad applicability in chemical applications of free energy simulations.

For the special, but commonly encountered case, in which $E(\mathbf{X}^N,\lambda)$ is linear in λ, as in

$$E(\mathbf{X}^N,\lambda) = (1-\lambda)E_0(\mathbf{X}^N) + \lambda E_1(\mathbf{X}^N) \tag{31}$$

$$E(\mathbf{X}^N,\lambda) = E_0(\mathbf{X}^N) + \lambda \Delta E(\mathbf{X}^N) \tag{32}$$

where

$$\Delta E(\mathbf{X}^N) = E_1(\mathbf{X}^N) - E_0(\mathbf{X}^N) \tag{33}$$

then

$$\frac{\partial E(\mathbf{X}^N,\lambda)}{\partial \lambda} = \Delta E(\mathbf{X}^N) \tag{34}$$

and

$$\Delta A = \int_0^1 < \Delta E(\mathbf{X}^N) >_\lambda d\lambda \tag{35}$$

Note that although $\Delta E(\mathbf{X}^N)$ is now independent of λ, the ensemble average must still be taken over the λ-distribution, $P(\mathbf{X}^N,\lambda)$. For the special case of $E_0(\mathbf{X}^N)$ being identically zero, the ideal gas reference state for the configurational energy,

$$\Delta A = \int_0^1 U(\lambda)\, d\lambda \tag{36}$$

where

$$U(\lambda) = \int \ldots \int E_1(\mathbf{X}^N)P(\mathbf{X}^N,\lambda)\, d\mathbf{X}^N \tag{37}$$

an internal energy-like quantity.

(B) Methodology

The methodology for carrying out a thermodynamic integration in a molecular simulation context is thus quite straightforward. A series of simulations is set up corresponding to a succession of discrete λ-values on the interval from 0 to 1, giving discrete values for the various $< \Delta E(\mathbf{X}^N) >_\lambda$. The auxiliary integration over λ is then carried out numerically. For an example, see Fig. 2. The λ-integration and the free energy determination thus requires typically 5-10 times as much computational effort than a mean energy calculation via molecular simulation, just to get one (albeit important!) number.

In favorable cases, it is possible to expedite things by allowing λ to vary continuously in the simulation, a procedure introduced by Berendsen et al. [18] and called 'slow growth'. Here, a single simulation is used to develop an estimate of $U(\lambda)$, with free energy obtained from a simple λ-integration at the end. Quite encouraging results have been obtained by this procedure on a picosecond time scale for simple systems. However, strictly speaking, the simulation has never properly equilibrated unless λ is infinitesimally small, and furthermore whatever structural changes occur in the system in going from initial to final state must

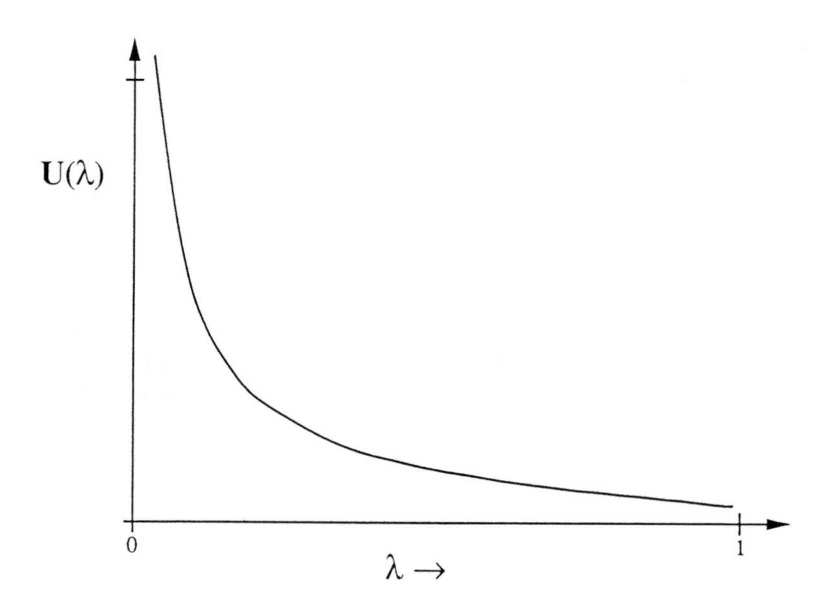

Fig. 2. The quantity $U(\lambda)$, i.e. $<\Delta E(X^N)>_\lambda$, vs. λ. The free energy computed via thermodynamic integration is the area under this curve. [From Swaminathan, S., Mezei, M. and Beveridge, D.L., J. Am. Chem. Soc., 100 (1978) 3255.]

be achieved on the picosecond time scale. Larger (and indeed, conformational) changes may not have had time to occur and could, in unfavorable cases, lead to problems with a slow growth as opposed to a discrete method wherein a thorough equilibration is conducted at each step. An insufficient sampling of configuration space of this type in a simulation is referred to as a 'quasi-ergodic' problem in molecular simulations.

Particular care must be taken with the thermodynamic integration calculation in the vicinity of $\lambda \rightarrow 0$ (or $\lambda \rightarrow 1$ as well for a mutation). Here, when the initial and final states put the atoms at different positions, the simulation tends to increasingly permit the configurations in $P(\mathbf{X}^N,\lambda)$ for which the actual ΔE is quite high due to clashes, leading to a rapid increase in $<\Delta E(\mathbf{X}^N)>_\lambda$ (possibly a singularity), and a concomitant loss of numerical accuracy, if not a complete blowup of the problem. A partial antidote to this is to employ an energy expression nonlinear in λ such as

$$E(\mathbf{X}^N,\lambda^m) = (1-\lambda^m)E_0(\mathbf{X}^N) + \lambda^m E_1(\mathbf{X}^N) \quad (38)$$

Here, as seen in Fig. 3, the energy approaches the asymptotes more sharply than in the linear case, building a knowledge of the coupled particle into the system at an earlier point in the integration. This problem requires considerable

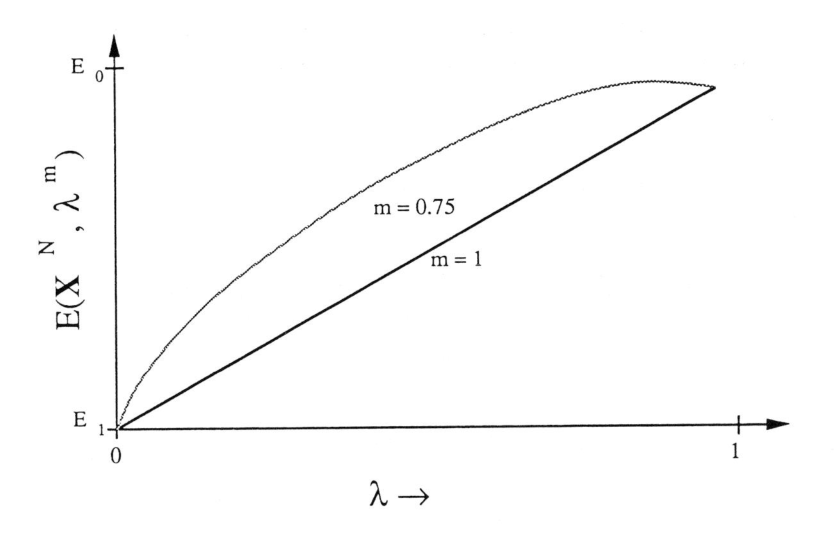

Fig. 3. Comparison of linear and non-linear dependence of configurational energy on λ (schematic).

practical attention when carrying out free energy simulations. Nonlinear methods have been discussed further by Mezei and Beveridge [12] and by Cross [19].

For examples of thermodynamic integration in molecular simulation, we find the early paper by Mruzik et al. [20] on a Monte Carlo study of ion water clusters to be instructive. The 'slow growth' procedure is described by Straatsma et al. [21] in a paper on the free energy of hydrophobic hydration, which is also recommended reading.

4. The perturbation method

(A) General formulation

The perturbation method for free energy simulations addresses the calculation of the free energy difference ΔA (or ΔG) for two states by an alternative route to that used in thermodynamic integration. The derivation however parallels that used earlier in the discussion of ensemble average expression for free energy in Eq. 13. We begin with the free energy difference written in the form of Eq. 19.

$$\Delta A = -kT \ln \frac{Z_1}{Z_0} \tag{39}$$

for an N-particle system in which the initial and final states differ by some topographical transformation. Inserting a unity in the form of

$$1 = \exp[+\beta E_0(\mathbf{X}^N)]\exp[-\beta E_0(\mathbf{X}^N)] \tag{40}$$

into the integrand of the partition function Z_1 in the numerator of Eq. 39, one obtains

$$\frac{Z_1}{Z_0} = \frac{\int\ldots\int \exp[-\beta E_1(\mathbf{X}^N)]\exp[+\beta E_0(\mathbf{X}^N)]\exp[-\beta E_0(\mathbf{X}^N)]\,d\mathbf{X}^N}{\int\ldots\int \exp[-\beta E_0(\mathbf{X}^N)]\,d\mathbf{X}^N} \tag{41}$$

$$\frac{Z_1}{Z_0} = \int\ldots\int \exp\{-\beta[E_1(\mathbf{X}^N) - E_0(\mathbf{X}^N)]\}\,P_0(\mathbf{X}^N)\,d\mathbf{X}^N \tag{42}$$

$$\frac{Z_1}{Z_0} = \langle\exp[-\beta\Delta E(\mathbf{X}^N)]\rangle_0 \tag{43}$$

where the subscript zero indicates configurational averaging over the ensemble of configurations representative of the initial, or reference state of the system. Then

$$\Delta A = -kT\ln\langle\exp[-\beta\Delta E(\mathbf{X}^N)]\rangle_0 \tag{44}$$

Also, by symmetry, since the initial and final states are in principle interchangeable,

$$\Delta A = -kT\ln\langle\exp[+\beta\Delta E(\mathbf{X}^N)]\rangle_1 \tag{45}$$

where the configurational averaging is performed with respect to the probability function representative of the final state of the system. Introduction of virtual particles into the coordinate list allows this treatment to be generalized to *any* change in the system.

This approach is numerically accurate only when the initial and final states of the system differ by a relatively small amount, so that one may be regarded as a 'perturbation' on the other. Thus, it is common practice in the chemical literature to refer to this approach as the 'perturbation method'. However, Eqs. 44 and 45 are in principle exact, provided there are no configurations for which $E_0(\mathbf{X}^N)$ is infinite but $E_1(\mathbf{X}^N)$ is finite. The earliest statement of these equations appears to be due to Zwanzig [22]. The use of Eqs. 44 and/or 45 to compute free energy is also known as the 'direct method' since in principle ΔA is obtained from a single simulation, as opposed to thermodynamic integration, which involves a series of simulations (or else, 'slow growth'), and a subsequent λ-quadrature to carry out the calculation.

(B) Methodology

The implementation of the perturbation method in its simplest manifestation involves defining $E_0(\mathbf{X}^N)$ and $E_1(\mathbf{X}^N)$ and running a Monte Carlo or molecular dynamics simulation for the 0th state of the system in the usual manner; that is, a sampling based on the probability function $P_0(\mathbf{X}^N)$, and forming the ensemble average of $\exp[-\beta\Delta E(\mathbf{X}^N)]$. Inserting into Eq. 44 gives the corresponding free energy difference. An analogous procedure can be formulated with respect to Eq. 45. These two calculations are referred to by practitioners as running 'forwards' or 'backwards', respectively, on the λ-coordinate. Better, but twice as much work, would be to calculate ΔA both ways and average the results. This is called 'double-ended' or 'double-wide' sampling [23]. The discrepancy between the two estimates of ΔA is an estimate in the statistical uncertainty in the result due to inaccuracies in the numerical integrations of Eqs. 44 and 45, and arise from configurations sampled in $P_0(\mathbf{X}^N)$ but not in $P_1(\mathbf{X}^N)$ and vice versa. It is essential to note that errors due to insufficient sampling in *both* directions, that is, important configurations missed in both $P_0(\mathbf{X}^N)$ and $P_1(\mathbf{X}^N)$ simulations, are not accounted for in this estimate of error; that is, it is an estimate of precision, but not necessarily of accuracy. This is another example of a quasi-ergodic problem, which along with potentials is a major source of inaccuracy in free energy determination by molecular simulation.

The discrepancy between results based on forward and backward integrations in the perturbation method is minimal when the initital and final states of the system are very similar. Then configurations important in the ensemble representative of the initital state are also important in the final state and vice versa, and are picked up in the sampling for both directions. Successful applications of Eqs. 44 and 45 are thus commonly limited to free energy differences $<2kT$, or ~1.5 kcal/mol. However, the free energy difference for many chemical and biomolecular processes is considerably larger than 2kT. The perturbation method can be used successfully in most cases by defining the intermediate states on the λ-coordinate which differ successively by no more than 2kT. The total free energy change can be obtained by summing the ($\leq 2kT$) ΔA's between the intermediate states,

$$\Delta A = \sum_{i=0}^{k-1} \Delta A(\lambda_i \rightarrow \lambda_{i+1}) \tag{46}$$

where the interval $\lambda:0$ to 1 has been divided up into k-subintervals. Obviously, the computation time increases with the number of subintervals involved since individual simulations are necessary for each intermediate state, λ_i.

The subinterval perturbation method can be performed as economically as possible using double-wide sampling [23], the current method of choice. This procedure is illustrated schematically in Fig. 4, for six subintervals. Consider an interior point, λ_i as typical, (cf. 'a' in Fig. 4) a molecular simulation is performed based on $P(\mathbf{X}^N,\lambda_i)$. Using $P(\mathbf{X}^N,\lambda_i)$ and $\Delta E(\lambda_i \rightarrow \lambda_{i+1})$, the free energy difference

$$\Delta A(\lambda_i \rightarrow \lambda_{i+1}) = -kT \ln \{ < \exp[-\beta \Delta E(\lambda_i \rightarrow \lambda_{i+1})] >_i \} \tag{47}$$

is evaluated, a forward step from λ_i (cf. 'b' in Fig. 4). Using the same simulation for $P(\mathbf{X}^N,\lambda_i)$ and $\Delta E(\lambda_i \rightarrow \lambda_{i-1})$, the difference

$$\Delta A(\lambda_i \rightarrow \lambda_{i-1}) = -kT \ln \{ < \exp[-\beta \Delta E(\lambda_i \rightarrow \lambda_{i-1})] >_i \} \tag{48}$$

is obtained; that is, the 'backwards' step in free energy from λ_i (cf. 'c' in Fig. 4). Thus the free energy difference for any two of these interior subintervals is obtained from a single simulation. Spanning the interval from $\lambda:0$ to 1 with k-subintervals, there will be k-2 interior points requiring (k-2)/2 independent simulations. Adding one on each end, where double-ended sampling is not

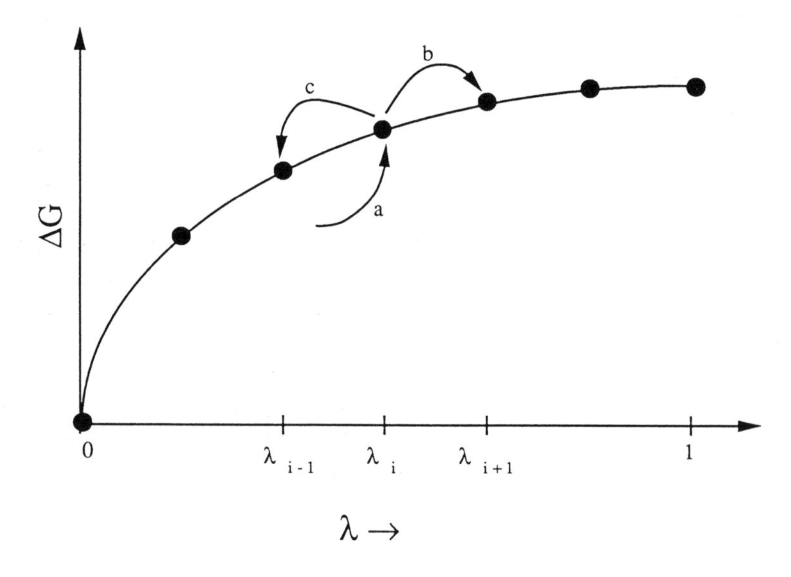

Fig. 4. Schematic illustration of double-wide sampling for use with the perturbation method.

possible, makes a total of $2+(k\text{-}2)/2$ simulations that are required to do k-subintervals between initial and final states by the perturbation method. The number of subintervals required for a given problem is established by trial with respect to the ~2kT criteria, and the type of problem that can be done is limited by the computer budget. Here, as in thermodynamic integration, the regions around $\lambda \to 0$ and $\lambda \to 1$ can be especially problematic, and sometimes it is necessary to make a finer grid in this region.

Instructive examples of the perturbation method used in conjunction with molecular simulation can be found in a calculation of the free energy of cavity formation in water by Postma et al. [24], and the calculation of the $\Delta\Delta G$ of hydration for the mutational change of $CH_3OH \to CH_3CH_3$ by Jorgensen and Ravimohan [25]. This latter case has become the classic test for diverse programs incorporating free energy methods into molecular simulation, and this rather complex calculation has been successfully repeated in about six different laboratories with different independently written computer programs.

5. The potential of mean force

(A) General formulation

For the special case of two of the N-particles of a system fixed in space at a distance, R, the expression for Helmholtz free energy takes the form

$$A(R) = -kT \ln \{ \exp[-E(\mathbf{X}^N \mid R)/kT] \} \, d\mathbf{X}^{N-2} \quad (49)$$

where $(\mathbf{X}^N \mid R)$ denotes a configuration conditional on the two fixed particles being R. The radial distribution function for the distance of any two particles is defined elsewhere [26] by the equation

$$g(R) = [N(N-1)/\rho] \exp[-E(\mathbf{X}^N \mid R)/kT] \, d\mathbf{X}^{N-2} \quad (50)$$

where ρ is the number density. Thus,

$$A(R) = -kT \ln[g(R)] + \text{constant} \quad (51)$$

The quantity A(R) is the reversible work involved in the association of the two particles from infinite separation to the distance R, in solution. The derivative of A(R) with respect to R is the force acting between the two particles; including both direct and solvent-averaged contributions. In the statistical mechanics literature, a quantity such as A(R) is frequently encountered and known as a potential of mean force, and usually denoted as w(R). The quantity R can be a simple distance, an internal coordinate representing a torsional degree of freedom, or a generalized structural alteration in the system. A knowledge of A(R) can be useful for the study of free energy due to conformational changes, molecular associations and the free energy profile for chemical reactions in condensed phases. A typical form for w(R) is shown in Fig. 5.

(B) Methodology

The potential of mean force can in principle be determined from simulation in two ways. The R-coordinate can be considered as an additional variable in the simulation, thus producing a direct estimate of g(R). As an alternative, it is possible to perform separate simulations at different fixed values of R, compute the forces due to the surroundings, and then integrate the force functions numerically. This latter route has been found to be numerically less reliable, and we focus herein on approaches involving, in one way or another, a variable R.

In a variable R simulation, it is important to note that g(R) is obtained as the ratio of the probability of sample the R-coordinate, and the volume element of the configuration space corresponding to the R-coordinate, viz.,

$$g(R) = P(R)/V(R) \tag{52}$$

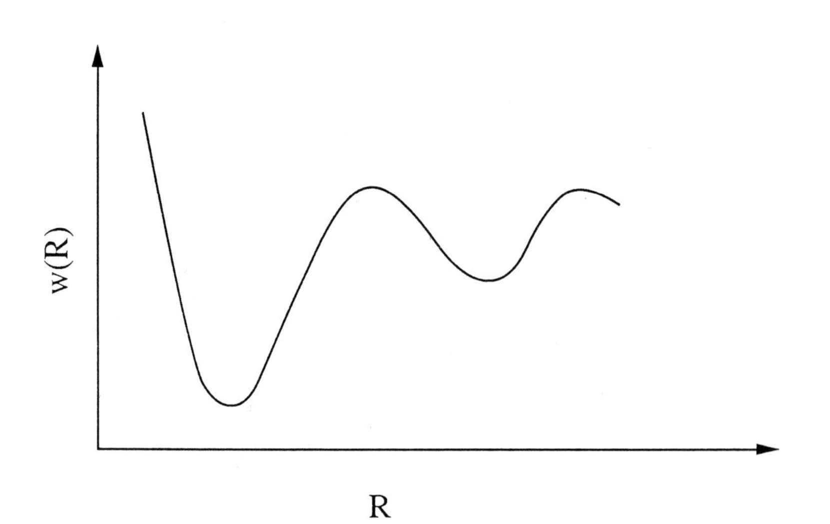

Fig. 5. Typical plot of potential of mean force w(R) vs. R for a simple molecular association. The oscillatory behavior, giving rise to contact and solvent-separated structures, is due to granularity in the solvent.

The volume element can be interpreted as a quantity proportional to the probability of sampling the parameter, R, with the potential function set to zero. For example, if R is an intermolecular distance, $V(R)=4\pi R^2$, and if R is a torsion angle, $V(R)=$ constant. Its determination becomes progressively more complex as the dimensionality of R is increased. Also, in general, subtle considerations of the Jacobian for the transformation as pointed out by Fixman [27], may need to be factored in.

In any case, when the R-coordinate is considered as simply another degree of freedom in an otherwise conventional mean energy calculation, serious sampling problems arise. In practical applications, w(R) varies for several kT in its range. This, however, converts into order-of-magnitude differences in g(R). The simulation, seeking to describe the equilibrium state dictated by Boltzmann statistics, would end up sampling only the most probable region of the R-space. As a result, all values of R would not be represented, and thus an accurate w(R) would not be obtained. However, the range of sampling can be extended systematically in a special form of importance sampling by running the Metropolis procedure or molecular dynamics with a modified energy function.

$$E'(\mathbf{X}^N) = E(\mathbf{X}^N) + W(\mathbf{X}^N) \tag{53}$$

Here, $W(\mathbf{X}^N)$ is a weighting function that can take various forms depending on the particular application, but typically is of the form

$$W_0(\mathbf{X}^N) = k_W(\mathbf{X}^N - \mathbf{X}_0{}^N)^2 \tag{54}$$

where k_w is an effective force constant. Values of $W(\mathbf{X}^N)$ are seen to go large for points in configuration space that is far removed from $\mathbf{X}_0$, and thus, a simulation based on the energy function, $E'(\mathbf{X}^N)$, will be biased towards the region of $\mathbf{X}_0$ to an extent depending on the value of the force constant, k_w. This technique is called 'umbrella sampling', since the sampling distributions are broader than the Boltzmann. The function $W_0(\mathbf{X}^N)$ is arbitrary, and using $E'(\mathbf{X}^N)$ rather than $E(\mathbf{X}^N)$ in a simulation, and generating umbrella distributions, carries the sampling to a non-Boltzmann regime. Thus, the ensemble averages obtained would not correspond properly to thermodynamic observables.

Valleau and Torrie [8], in research that has come to have broad, important implications in free energy simulations, devised a procedure for extracting proper Boltzmann-weighted ensemble averages from a non-Boltzmann sampling. Consider the ensemble average $<Q>$ of any property, Q,

$$<Q> = \frac{\int \ldots \int Q(\mathbf{X}^N) \exp[-\beta E(\mathbf{X}^N)]\, d\mathbf{X}^N}{\int \ldots \int \exp[-\beta E(\mathbf{X}^N)]\, d\mathbf{X}^N} \tag{55}$$

Inserting unity in the form of

$$1 = \exp[+\beta W(\mathbf{X}^N)]\exp[-\beta W(\mathbf{X}^N)] \tag{56}$$

one obtains, since the exponential of a sum is the product of exponentials of the individual terms in the summation,

$$<Q> = \frac{\int \ldots \int Q(\mathbf{X}^N)\exp[-\beta E(\mathbf{X}^N)]\exp[+\beta W(\mathbf{X}^N)]\exp[-\beta W(\mathbf{X}^N)]\,d\mathbf{X}^N}{\int \ldots \int \exp[-\beta E(\mathbf{X}^N)]\exp[+\beta W(\mathbf{X}^N)]\exp[-\beta W(\mathbf{X}^N)]\,d\mathbf{X}^N} \tag{57}$$

Division of both numerator and denominator of Eq. 57 by

$$\int \ldots \int \exp[-\beta E'(\mathbf{X}^N)]\,d\mathbf{X}^N \tag{58}$$

yields

$$<Q> = \frac{\int \ldots \int Q(\mathbf{X}^N)\exp[\beta W(\mathbf{X}^N)]\,P_W(\mathbf{X}^N)\,d\mathbf{X}^N}{\int \ldots \int \exp[\beta W(\mathbf{X}^N)]\,P_W(\mathbf{X}^N)\,d\mathbf{X}^N} \tag{59}$$

where

$$P_W(\mathbf{X}^N) = \frac{\exp[-\beta E'(\mathbf{X}^N)]}{\int \ldots \int \exp[-\beta E'(\mathbf{X}^N)]\,d\mathbf{X}^N} \tag{60}$$

A more concise notation for Eq. 59 is

$$<Q> = \frac{<Q(\mathbf{X}^N)\exp[\beta W(\mathbf{X}^N)]>_W}{<\exp[\beta W(\mathbf{X}^N)]>_W} \tag{61}$$

where the subscript, W, indicates an ensemble average based on the probability, $P_W(\mathbf{X}^N)$ and the energy function, $E'(\mathbf{X}^N)$. Thus, obtaining a proper Boltzmann-referenced ensemble average $<Q>$ from a simulation which involves non-Boltzmann sampling involves forming auxiliary ensemble averages indicated in Eq. 61 over the probability $P_W(\mathbf{X}^N)$, and forming the quotient indicated.

One obvious limitation of the umbrella sampling procedure is the following: the more the sampling is to be extended, the larger will be the range of $W(\mathbf{X}^N)$, making the calculation of $<\exp[\beta W(\mathbf{X}^N)]>_W$ more prone to numerical uncertainties. In fact, if the variations are especially large, the computed averages will be dominated by only a very few terms, a clearly undesirable effect. Therefore, a limit must be set on the variations permitted by $W_0(\mathbf{X}^N)$ which effectively puts a limit on the extension of sampling by this procedure. A succession of simulations using first $W_0(\mathbf{X}^N)$, and then $W_1(\mathbf{X}^N)$, $W_2(\mathbf{X}^N)$, ... can be devised to overcome this problem [28]. The complete w(R) for all R can then be constructed by a matching of individual probability functions for which considerable overlap is observed. Any two successive distributions k and l do not superimpose absolutely in the overlap region because the normalization constants for the individual distributions, P_k and P_l, depend in a non-trivial way on the window. However, P_l can be renormalized to obtain a new $P'_l(r)$,

$$P'_l(r) = P_l(r)N_{lk} \tag{62}$$

such that k and l are referred to the same normalization factor. A common choice has been simply

$$N_{lk} = P_k(R_\alpha)/P_l(R_\alpha) \tag{63}$$

where R_α is, in principle, any point a where $P_k$and P_l overlap. The statistics for the $P'_k(r)$ get poorer towards the tails of the distributions, and thus, it is essential to match the distribution at a point R_α where both P_k and P_l have simultaneously reasonable error bars. The final function should not, however, depend sensitively on where the matching is carried out. Ideally, the matching should be based on all overlapping points, with higher weight given to points that were sampled more extensively. A formalism for this, applicable to multidimensional matchings, has been suggested, particularly by Mezei et al. [29]. With this approach, a potential of mean force, w(R) can be developed in a series of molecular simulations for a considerable range of R. The idea of extracting ΔA from the end points of a potential of mean force determination is the basis for a free energy technique called the probability ratio method [29].

Particular important and instructive potential of mean force calculations have been described by Pangali et al. [28] in a Monte Carlo study of the hydrophobic interaction, and by Chandrasekar et al. [30] on the determination of free energy profile for an S_N2 bimolecular nucleophilic substitution reaction in gas phase and water. The series of papers dealing with the potential of mean force for *n*-butane in various solvents [31] provides informative detail on the application of this procedure to a conformational coordinate. The probability ratio method has been applied to the problem of solvent effects on the conformation stability of a dipeptide in water by Mezei et al. [29].

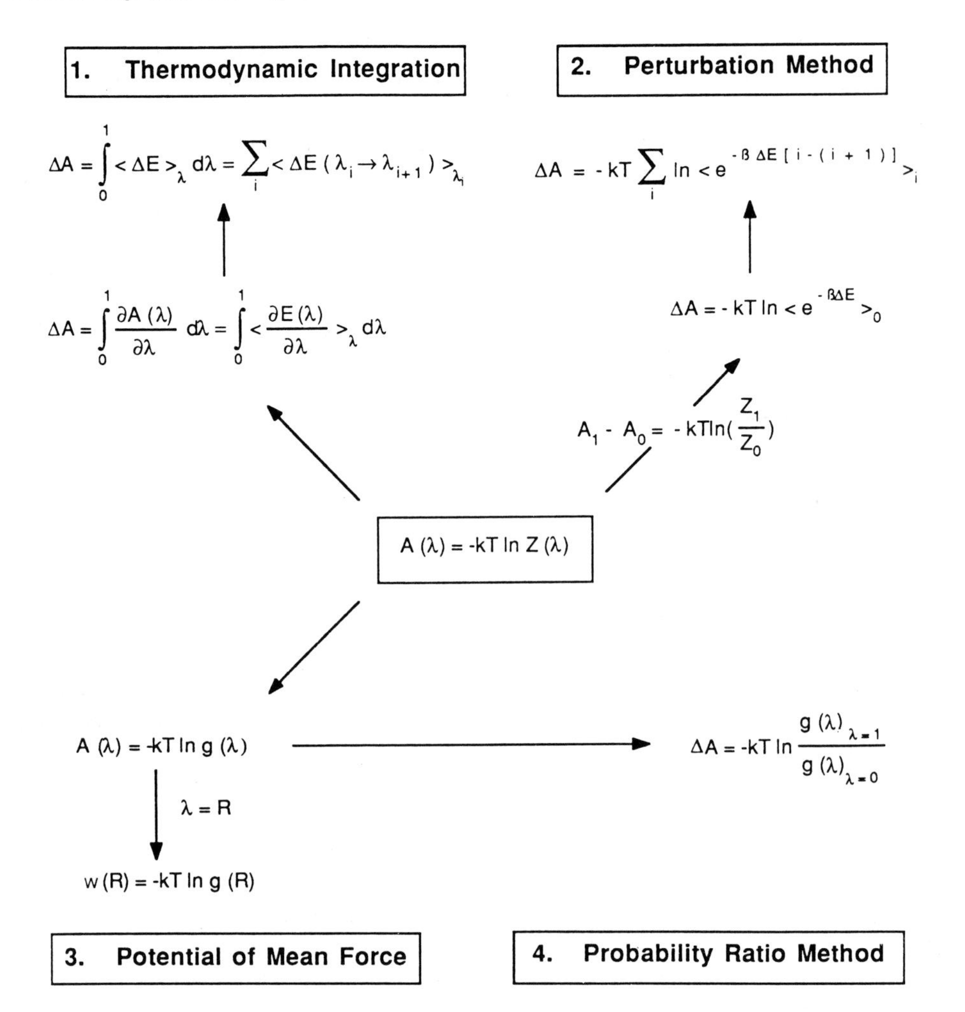

Fig. 6. Derivation of expressions used in free energy determinations via molecular simulation.

6. Summary and Conclusions

In this article, we have attempted to take the problem of free energy determination via molecular simulation, and present, at the most elementary level possible, the fundamental issues involved, the theoretical basis for several of the currently most popular methods for free energy simulation: thermodynamic integration, the perturbation method and the potential of mean force. A summary diagram of the relationship between the various free energy formulations is given in Fig. 6. In each case, the methodologies involved in implementing the free energy calculations in molecular simulations have been reviewed. Capabilities for carrying out free energy determinations are now implemented in each of the major distributed simulation programs, including GROMOS 86, AMBER 3.0 and CHARMm.

The free energy methods have recently been used in novel ways in conjunction with thermodynamic cycles to treat relative free energies of substrate binding to macromolecules [32,33], an application with broad implications in biochemical and pharmaceutical applications. However, it should be recognized that the *free energy methods do not overcome the multiple minima problem* in applications to macromolecules; this can be considered another manifestation of the quasi-ergodic difficulties mentioned earlier. However, one should note a recent paper by Scheraga et al. [34] who apply a variant of free energy methodology called adaptive importance sampling (see also Mezei [35]) to the multiple minima problem with encouraging results.

Due to the fundamental importance of a knowledge of free energy for the study of a molecular system, we expect that calculations of free energy via molecular simulation will become increasingly popular in the future. The disadvantage is the inordinate computation time involved, but combined with fast computers, and well-informed project design and structural and dynamical analysis results, free energy simulations can provide good numbers, useful insights, and forge a vital link between theoretical and computational science and experiment.

Acknowledgements

This article is in part derived from and in part builds upon a previous review by Mezei and Beveridge [12]. Previous conversations and collaborations with Dr. M. Mezei on this subject are gratefully acknowledged.

Research in this laboratory is supported by grants from the National Institutes of Health (GM-37909), National Science Foundation (CHE-8696117), the U.S. Office of Naval Research, and contributions from Merck, Sharpe and Dohme Research Laboratories, Bristol Meyers Inc, and the National Bureau of Standards.

References

1. Metropolis, N.A., Rosenbluth, A.W., Rosenbluth, M.N., Teller, A.H. and Teller, E.J., J. Chem. Phys., 21 (1953) 1987.
2. Alder, B.J. and Wainwright, T.E., J. Chem. Phys., 33 (1960) 1439.
3. Rahman, A. and Stillinger, F.H., J. Chem. Phys., 55 (1971) 3336.
4. Allen, M.P. and Tildesley, D.J., Computer Simulation of Liquids, Clarendon Press, Oxford, 1987.

5. McCammon, J.A. and Harvey, S.C., Dynamics of Proteins and Nucleic Acids, Cambridge University Press, Cambridge, 1987.
6. Hagler, A.T., Storm, P.S., Sharon, R., Becker, J.M. and Nasder, F.J., J. Am. Chem. Soc., 101 (1979) 6842.
7. Karplus, M. and Kushick, J., Macromolecules, 14 (1981) 325.
8. Valleau, J.P. and Torrie, G.M., In Berne, B.J. (Ed.) Statistical Mechanics, Part A: Equilibrium Techniques (Modern Theoretical Chemistry, Vol. 5), Plenum Press, New York, NY, 1977, Ch. 4 and 5.
9. Quirke, N., Proceedings of the NATO Summerschool on Superionic Conductors, Odense, Denmark, Plenum Press, New York, NY, 1980.
10. Shing, K.S. and Gubbins, K.E., In Haile, J.M. and Mansoori, G.A. (Eds.) Molecular-based Study of Fluids (Advances in Chemistry, vol. 204), American Chemical Society, Washington, DC, 1983.
11. Pohorille, A. and Pratt, L.R., In Packeri, L. (Ed.) Biomembranes (Methods in Enzymology, Vol. 127), Academic Press, New York, NY, 1986.
12. Mezei, M. and Beveridge, D.L., Ann. N.Y. Acad. Sci., 482 (1986) 1.
13. Barker, J.A. and Henderson, D., Rev. Mod. Phys., 48 (1976) 587.
14. Wood, W.W., In Temperly, H.N.V., Rowlinson, J.S. and Rushbrooke, G.S. (Eds.) Physics of Simple Liquids, North-Holland Publishing Company, Amsterdam, 1968.
15. Straatsma, T.P., Berendsen, H.J.C. and Stam, A.J., Mol. Phys., 57 (1986) 89.
16. Kirkwood, J.G., In Alder, B.J. (Ed.) Theory of Liquids, Gordon and Breach, New York, NY, 1968.
17. De Donder, Th., L'Affinité, Gauthier-Villars, Paris, 1927 / De Donder, Th. and van Rysselberche, P., Affinity, Stanford University Press, Stanford, CA, 1936.
18. Berendsen, H.J.C., Postma, J.P.M. and van Gunsteren, W.F., In Hermans, J. (Ed.) Molecular Dynamics and Protein Structure, Polycrystal Book Service, P.O. Box 27, Western Springs, IL 60558, 1985.
19. Cross, A.J., Ann. New York Acad. Sci., 482 (1986) 89.
20. Mruzik, M.R., Abraham, F.F., Schreiber, D.E. and Pound, G.M., J. Chem. Phys., 64 (1976) 481.
21. Straatsma, T.P., Berendsen, H.J.C. and Postma, J.P.M., J. Chem. Phys., 85 (1986) 6720.
22. Zwanzig, R.W., J. Chem. Phys., 22 (1954) 1420.
23. Bennet, C.H., J. Comp. Phys., 22 (1976) 245.
24. Postma, J.P.M., Berendsen, H.J.C. and Haak, J.R., Faraday Symp. Chem. Soc., 17 (1982) 55.
25. Jorgensen, W.L. and Ravimohan, C., J. Chem. Phys., 83 (1985) 3050.
26. Ben Naim, A., Water and Aqueous Solutions, Plenum Press, New York, NY, 1974.
27. Fixman, M., Proc. Natl. Acad. Sci. U.S.A., 71 (1974) 3050.
28. Pangali, C.S., Rao, M. and Berne, B.J., J. Chem. Phys., 71 (1979) 2975.
29. Mezei, M., Mehrota, P.K. and Beveridge, D.L., J. Am. Chem. Soc., 107 (1985) 2239.
30. Chandrasekar, J., Smith, S.F. and Jorgensen, W.L., J. Am. Chem. Soc., 106 (1985) 3049 and 107 (1985) 154.
31. Jorgensen, W.L., J. Phys. Chem., 87 (1983) 5304.
32. Tembe, B.L. and McCammon, J.A., Comput. Chem., 8 (1984) 281.
33. Van Gunsteren, W.F. and Berendsen, H.J.C., J. Comput.-Aided Mol. Design, 1 (1987) 171.
34. Scheraga, H.A. and Paine, G.H., Ann. New York Acad. Sci., 482 (1986) 60.
35. Mezei, M., J. Comp. Phys., 68(1987) 237.

Methods for calculation of free energies and binding constants: Successes and problems

W.F. van Gunsteren
Department of Physical Chemistry, University of Groningen,
Nijenborgh 16, 9747 AG Groningen, and
Department of Physics, Free University,
P.O. Box 7161, 1007 MC Amsterdam, The Netherlands

Summary

Methods for the calculation of free energy are briefly discussed. One of these, the so-called thermodynamic cycle integration technique, which combines well-known results from statistical thermodynamics with powerful computer simulation methods, is very promising with respect to practical applications in chemistry. Its basic formulae for the free energy difference between two equilibrium states A and B of a system are derived. An expression for the entropy difference is also given. A formula by which the free energy difference along a specified coordinate d between two non-equilibrium states of the system characterized by the values d_A and d_B of d can be calculated, is derived. Finally, a list of assumptions, approximations, practical considerations and limitations concerning methods for the computation of free energy is given and discussed.

1. Introduction

During the past two decades, molecular dynamics (MD) computer simulations have considerably added to our understanding at the atomic level of the properties of molecular systems such as liquids or solutions. A rather static picture of molecular conformation has gradually been transformed into a more dynamic one, where the molecular properties are dynamic averages over an ensemble of molecular configurations. The development of computer simulation techniques has been made possible by the continuous and rapid development of computer hardware. Every six to seven years the ratio of performance to price has increased by a factor of ten and due to the emerging parallel computing techniques the end of this increase is not yet in sight. Presently, small macromolecules, like proteins, in aqueous solution involving many thousands of atoms can be simulated over periods of about 10-100 ps. For reviews we refer to Refs. 1-6.

From a molecular dynamics trajectory the statistical equilibrium averages can be obtained for any desired directly measurable property of the system, that is, for which the value can be computed at each point of the trajectory. Examples of such properties are the kinetic energy of relevant parts of the system, structural properties, electric fields, etc. From such averages a number of thermodynamic properties can be derived. However, two important thermodynamic quantities, the entropy and the (Helmholtz or Gibbs) free energy can generally not be derived from a statistical average. These are global properties that depend on the extent of phase (or configuration) space accessible to the molecular system. Therefore, computation of the absolute free energy of a molecular system is virtually impossible. Yet, the most interesting chemical quantities like binding constants of donor-acceptor complexes or molecular solubilities are directly related to the free energy. Over the past few years several statistical mechanical procedures have evolved for evaluating relative free energies [7-9]. They are rather demanding as far as computer time is concerned, but will open up a wide area of most interesting applications in chemistry.

2. Methods to determine free energy or entropy

In a canonical ensemble the fundamental formula for the Helmholtz *free energy* F is [10]

$$F = -k_B T \ln Z \tag{2.1}$$

where k_B is Boltzmann's constant, T refers to the absolute temperature and Z is the partition function, determined by the Hamiltonian $H(\mathbf{p},\mathbf{q})$ that describes the total energy of the system in terms of (generalized) coordinates **q** and momenta **p** of the atoms in the system. For a system consisting of N atoms one has

$$Z = [h^{3N} N!]^{-1} \iint \exp[-H(\mathbf{p},\mathbf{q})/k_B T] \, d\mathbf{p} d\mathbf{q} \tag{2.2}$$

where h is Planck's constant. The factor N! is present when the N atoms are indistinguishable. When the atoms are distinguishable, such as atoms in a macromolecule, the factorial should not be present in (2.2). The *entropy* S is related to the free energy by

$$F = E - TS \tag{2.3}$$

where E is the total (kinetic plus potential) energy, the ensemble average $\langle H \rangle$ of $H(\mathbf{p},\mathbf{q})$. In terms of the *phase space probability*

$$\pi(\mathbf{p},\mathbf{q}) = \frac{\exp[-H(\mathbf{p},\mathbf{q})/k_B T]}{[h^{3N} N! \, Z]} \tag{2.4}$$

one has

$$E = \langle H \rangle = \iint H(\mathbf{p},\mathbf{q})\ \pi(\mathbf{p},\mathbf{q})\ d\mathbf{p}d\mathbf{q} \qquad (2.5)$$

Using (2.1-5) the entropy S can be expressed in terms of the probabilities $\pi(\mathbf{p},\mathbf{q})$ in phase space,

$$S = -k_B \iint \pi(\mathbf{p},\mathbf{q}) \ln[\ \pi(\mathbf{p},\mathbf{q}) h^{3N} N!\ ]\ d\mathbf{p}d\mathbf{q} \qquad (2.6)$$

In order to discuss the various techniques for evaluating F or S we must distinguish between diffusive and non-diffusive systems [7].

Diffusive systems, such as liquids or solutions, are characterized by the eventual diffusion of atoms or molecules over all of the available space. *Non-diffusive systems*, such as solids, glasses and macromolecules with a definite structure are characterized by time-independent average positions around which the atoms are fluctuating. For these systems the configuration space is limited and the integral (2.6) can be obtained by a direct determination of the probability distribution function $\pi(\mathbf{p},\mathbf{q})$, e.g., from a computer simulation. For diffusive systems this road is closed, since the multi-dimensional configuration space is so vast that it can never be integrated using simulation techniques. For this type of system two other techniques are available, probe methods and the very powerful thermodynamic perturbation and integration methods.

Methods for *direct determination of the probability distribution* are given in Refs. 11-13. If the potential energy function is not dependent on the momenta $\mathbf{p}$ of the atoms, the kinetic contribution to the entropy can be split off and analytically determined. The configurational part S_c depends on the multi-dimensional probability distribution $\pi(\mathbf{q})$, which is difficult to determine with sufficient accuracy from a simulation. However, if it is assumed to be a multivariate Gaussian distribution, it is completely characterized by its covariance matrix

$$\sigma_{ij} = \langle [q_i - \langle q_i \rangle][q_j - \langle q_j \rangle] \rangle \qquad (2.7)$$

and the configurational entropy becomes only a function of the determinant of the covariance matrix

$$S_c = \frac{1}{2} k_B [\, 3N\,(1 + \ln(2\pi)) + \ln(\det \underline{\sigma})\,] \qquad (2.8)$$

When treating a macromolecule, the computation of this determinant of a matrix of thousands of degrees of freedom is quite computer time consuming. The application of this method is also restricted by the fact that the inclusion of diffusive solvent degrees of freedom is not possible.

Probe methods [14-17], like the particle insertion methods, insert a test atom in the system at a very large number of uniformly distributed random test positions. The interaction energy of the test atom with the atoms in the system is computed and the average Boltzmann factor of the obtained energy yields the free energy. These methods work well for homogeneous low and medium density systems. At high density or in heterogeneous systems poor statistics are obtained when averaging the Boltzmann factor, due to the generally large insertion energies.

Thermodynamic perturbation and integration methods will be described in Section 4.

3. Simulation methods for molecular systems

When simulating a molecular system, an *energy function* or interaction potential or *force field* has to be postulated, which describes the potential energy of the molecular system as a function of the (Cartesian) positions $\mathbf{r}_i$ of the N atoms labeled by the index i. A typical example looks as follows [18]

$$V(\mathbf{r}) \equiv V(\mathbf{r}_1,\mathbf{r}_2,\ldots,\mathbf{r}_N) =$$

$$\sum_{\text{bonds}} \tfrac{1}{2} K_b [b-b_0]^2 + \sum_{\text{angles}} \tfrac{1}{2} K_\theta [\theta-\theta_0]^2 + \sum_{\text{torsions}} \tfrac{1}{2} K_\zeta [\zeta-\zeta_0]^2 + \tag{3.1}$$

$$\sum_{\text{dihedrals}} K_\phi [1+\cos(n\phi-\delta)] + \sum_{\text{pairs (i,j)}} \left[\frac{C_{12}(i,j)}{r_{ij}^{12}} - \frac{C_6(i,j)}{r_{ij}^{6}} + \frac{q_i q_j}{(4\pi\epsilon_0\epsilon_r r_{ij})} \right]$$

The first term represents the covalent bond-stretching interaction along bond b. It is a harmonic potential in which the minimum energy bond length b_0 and the force constant K_b vary with the particular type of bond. The second term describes the bond-angle bending interaction in a similar form. Two forms are used for the torsional or dihedral-angle interactions: a harmonic term for dihedrals (torsions ζ) that are not allowed to make transitions, e.g. dihedral angles within aromatic rings, and a sinusoidal term for the other dihedrals (ϕ), which may make 360-degree turns. The last term in (3.1) is a sum over all pairs of atoms and represents the effective nonbonded interaction, composed of the van der Waals and the Coulomb interaction between atoms i and j with charges q_i and q_j at a distance r_{ij}.

The next step is to choose a method by which an *ensemble of configurations* of the molecular system can be generated. We mention four techniques.

(1) If the system contains only a few degrees of freedom, the complete

configuration space can be scanned for low energy molecular configurations and an ensemble can be generated using the Boltzmann factor.

$$\exp[-V(\mathbf{r}_1,\mathbf{r}_2,\ldots,\mathbf{r}_N)/k_BT] \tag{3.2}$$

as a weight function. This method is called the *systematic search* method (SS).

(2) If a system contains many degrees of freedom, straightforward scanning of the complete configuration space is impossible. In that case, an ensemble of configurations can be generated by the *Monte Carlo* (MC) method, which employs a combination of random sampling and use of the Boltzmann factor (3.2). Given a starting configuration a new configuration is generated by random displacement of one (or more) atoms. The displacements should be such that in the limit of a large number of successive displacements the available Cartesian space of all atoms is uniformly sampled. The newly generated configuration is either accepted or rejected on the basis of an energy criterion involving the change ΔV of the potential energy (3.1) with respect to the previous configuration. In the case of rejection, the previous configuration is counted again and used as a starting point for another random displacement. The criterion is the following: accept if $\Delta V \leq 0$, or for $\Delta V > 0$, accept if $\exp(-\Delta V/kT) > R$, where R is a random number taken from a uniform distribution over the interval (0,1). In this way each configuration occurs with a probability proportional to its Boltzmann factor (3.2). In order to obtain high computational efficiency, one would like to combine a large (random) step size with a high acceptance ratio. For complex systems involving many covalently bound atoms, a reasonable acceptance ratio can only be obtained for a very small step size. This makes MC much less efficient than MD for (macro)molecular systems.

(3) By application of the simulation method of *molecular dynamics* (MD) a trajectory (configurations as a function of time) of the system is generated by simultaneous integration of Newton's equations of motion for all atoms $(i=1,2,\ldots,N)$ in the system

$$\frac{d^2\mathbf{r}_i(t)}{dt^2} = m_i^{-1}\,\mathbf{F}_i(t) \tag{3.3}$$

where the force exerted on atom i with mass m_i is found from

$$\mathbf{F}_i(t) = -\frac{\partial V(\mathbf{r}_1(t),\mathbf{r}_2(t),\ldots,\mathbf{r}_N(t))}{\partial \mathbf{r}_i(t)} \tag{3.4}$$

(4) An extension of the method of MD is the technique of *stochastic dynamics*

(SD). A trajectory of the system is generated by integration of the stochastic Langevin equation of motion.

$$\frac{d^2\mathbf{r}_i(t)}{dt^2} = m_i^{-1}\,[\,\mathbf{F}_i(t)+\mathbf{R}_i(t)\,]-\gamma_i\frac{d\mathbf{r}_i(t)}{dt} \tag{3.5}$$

Here, a stochastic force $\mathbf{R}_i$ and a frictional force proportional to a friction coefficient γ_i have been added to (3.3).

4. Thermodynamic perturbation and integration methods

Thermodynamic perturbation and integration methods [19-37] make use of the fact that the free energy changes related to small perturbations of a molecular system can be determined from a simulation. The free energy difference between two states A and B of a system can be determined from a MD simulation in which the potential energy function V($\mathbf{r}$) (3.1) is slowly changed such that the system slowly changes from state A to state B over a reversible path.

The method works as follows. First, the Hamiltonian $H(\mathbf{p},\mathbf{r})$, or only the potential energy term $V(\mathbf{r})$, is made a function of a *coupling parameter* λ, such that $H(\mathbf{p},\mathbf{r},\lambda_A)$ characterises state A of the system and $H(\mathbf{p},\mathbf{r},\lambda_B)$ state B. For example

$$H(\mathbf{p},\mathbf{r},\lambda) = \sum_{i=1}^{N}\frac{\mathbf{p}_i^2}{2m_i(\lambda)} + V(\mathbf{r}_1,\mathbf{r}_2,\ldots,\mathbf{r}_N,\lambda) \tag{4.1}$$

with

$$m_i(\lambda) = [\,1-\lambda\,]\,m_i^A+\lambda\,m_i^B \tag{4.2}$$

and

$$V(\mathbf{r}_1,\mathbf{r}_2,\ldots,\mathbf{r}_N,\lambda) =$$

$$\sum_{\text{bonds } b} V(b,\lambda) + \sum_{\text{angles } \theta} V(\theta,\lambda) + \sum_{\text{torsions } \zeta} V(\zeta,\lambda) + \sum_{\text{dihedrals } \phi} V(\phi,\lambda) + \sum_{\text{pairs}(i,j)} V(r_{ij},\lambda) \tag{4.3}$$

where

$$V(b,\lambda) = \frac{1}{2}\left\{[\,1-\lambda\,]K_b^A+\lambda K_b^B\right\}\left\{b-\{[\,1-\lambda\,]b_0^A+\lambda b_0^B\}\right\}^2 \tag{4.4}$$

$$V(\theta,\lambda) = \frac{1}{2}\left\{[1-\lambda]K_\theta^A + \lambda K_\theta^B\right\} \left\{\theta - \{[1-\lambda]\theta_0^A + \lambda\theta_0^B\}\right\}^2 \tag{4.5}$$

$$V(\zeta,\lambda) = \frac{1}{2}\left\{[1-\lambda]K_\zeta^A + \lambda K_\zeta^B\right\} \left\{\zeta - \{[1-\lambda]\zeta_0^A + \lambda\zeta_0^B\}\right\}^2 \tag{4.6}$$

$$V(\phi,\lambda) = [1-\lambda]K_\phi^A[1+\cos(n^A\phi - \delta^A)] + \lambda K_\phi^B[1+\cos(n^B\phi - \delta^B)] \tag{4.7}$$

$$\begin{aligned} V(r_{ij},\lambda) = {} & [1-\lambda]\left\{\frac{C_{12}^A(i,j)}{r_{ij}^{12}} - \frac{C_6^A(i,j)}{r_{ij}^6} + \frac{q_i^A q_j^A}{(4\pi\epsilon_0 r_{ij})}\right\} + \\ & \lambda\left\{\frac{C_{12}^B(i,j)}{r_{ij}^{12}} - \frac{C_6^B(i,j)}{r_{ij}^6} + \frac{q_i^B q_j^B}{(4\pi\epsilon_0 r_{ij})}\right\} \end{aligned} \tag{4.8}$$

The Helmholtz free energy of the system becomes a function of λ:

$$F(\lambda) = -k_B T \ln Z(\lambda) \tag{4.9}$$

and so does the partition function

$$Z(\lambda) = [h^{3N}N!]^{-1} \iint \exp[-H(\mathbf{p},\mathbf{r},\lambda)/k_B T]\, d\mathbf{p}d\mathbf{r} \tag{4.10}$$

The free energy difference ΔF_{BA} then reads

$$\Delta F_{BA} = F(\lambda_B) - F(\lambda_A) = -k_B T \ln \frac{Z(\lambda_B)}{Z(\lambda_A)} \tag{4.11}$$

which can be expressed as an ensemble average

$$\begin{aligned} \Delta F_{BA} = {} & \\ & -k_B T \ln \left\{ \frac{\iint \exp[-(H(\mathbf{p},\mathbf{r},\lambda_B) - H(\mathbf{p},\mathbf{r},\lambda_A))/k_B T]\exp[-H(\mathbf{p},\mathbf{r},\lambda_A)/k_B T]\, d\mathbf{p}d\mathbf{r}}{\iint \exp[-H(\mathbf{p},\mathbf{r},\lambda_A)/k_B T]\, d\mathbf{p}d\mathbf{r}} \right\} \\ = {} & -k_B T \ln\{\langle \exp[-(H(\mathbf{p},\mathbf{r},\lambda_B) - H(\mathbf{p},\mathbf{r},\lambda_A))/k_B T]\rangle_{\lambda_A}\} \end{aligned} \tag{4.12}$$

where the brackets $<...>_\lambda$ mean an ensemble average over the coordinates $\mathbf{r}$ and momenta $\mathbf{p}$ at the value λ. Formula (4.12) is called the *perturbation formula*, since it will only yield accurate results when state B is close to state A. If this difference is large, the change from A to B must be split up in a number of steps between intermediate states that are close enough to allow for the use of formula (4.12) and then ΔF_{BA} is just the sum of the ΔF for all intermediate steps.

Straightforward differentiation of (4.9) with respect to λ at constant T yields

$$\left[\frac{\partial F(\lambda)}{\partial \lambda}\right]_T = - \frac{k_B T}{Z(\lambda)} \left[\frac{\partial Z(\lambda)}{\partial \lambda}\right]_T$$

$$= \frac{\iint \frac{\partial H(\mathbf{p},\mathbf{r},\lambda)}{\partial \lambda} \exp[-H(\mathbf{p},\mathbf{r},\lambda)/k_B T]\, d\mathbf{p} d\mathbf{r}}{\iint \exp[-H(\mathbf{p},\mathbf{r},\lambda)/k_B T]\, d\mathbf{p} d\mathbf{r}}$$

$$= <\frac{\partial H(\mathbf{p},\mathbf{r},\lambda)}{\partial \lambda}>_\lambda \tag{4.13}$$

leading to the *integration formula* for the free energy difference ΔF_{BA}:

$$\Delta F_{BA} = \int_{\lambda_A}^{\lambda_B} <\frac{\partial H(\mathbf{p},\mathbf{r},\lambda)}{\partial \lambda}>_\lambda d\lambda \tag{4.14}$$

If λ is being changed very slowly from λ_A to λ_B during a MD simulation, the integration (4.14) can be carried out in the course of the MD run. In this way ΔF_{BA} can be obtained for rather different states A and B, as long as the continuous change in λ is so slow that the system remains essentially in equilibrium for each intermediate value of λ.

The corresponding formulae for the isobaric ensemble can be found in Ref. 34.

5. Thermodynamic integration formula for the entropy

A thermodynamic integration formula for the entropy difference ΔS_{BA} between two states A and B of the system can be derived along the same lines as was done in the previous section for the free energy difference ΔF_{BA}.

The entropy can be found [10] by differentiating (2.3) with respect to T (at constant N and volume V),

$$S = - \frac{\partial F}{\partial T} \tag{5.1}$$

Using (2.1) and (2.2) one finds

$$S(\lambda) = \tag{5.2}$$

$$+ k_B \ln Z(\lambda) + [T\, Z(\lambda)]^{-1}[h^{3N} N!]^{-1} \iint H(\mathbf{p},\mathbf{r},\lambda) \exp[-H(\mathbf{p},\mathbf{r},\lambda)/k_B T]\, d\mathbf{p} d\mathbf{r}$$

Straightforward differentiation of this formula with respect to λ at constant T yields

$$\left[\frac{\partial S(\lambda)}{\partial \lambda}\right]_T =$$

$$[k_B T^2]^{-1} \left[<H(\mathbf{p},\mathbf{r},\lambda)>_\lambda < \frac{\partial H(\mathbf{p},\mathbf{r},\lambda)}{\partial \lambda} >_\lambda - < H(\mathbf{p},\mathbf{r},\lambda) \frac{\partial H(\mathbf{p},\mathbf{r},\lambda)}{\partial \lambda} >_\lambda \right] \tag{5.3}$$

leading to the *integration formula for the entropy difference* ΔS_{BA}:

$$T\, \Delta S_{BA} = T\,[S(\lambda_B) - S(\lambda_A)] =$$

$$[k_B T]^{-1} \int_{\lambda_A}^{\lambda_B} \left[<H(\mathbf{p},\mathbf{r},\lambda)>_\lambda < \frac{\partial H(\mathbf{p},\mathbf{r},\lambda)}{\partial \lambda} >_\lambda - < H(\mathbf{p},\mathbf{r},\lambda) \frac{\partial H(\mathbf{p},\mathbf{r},\lambda)}{\partial \lambda} >_\lambda \right] d\lambda \tag{5.4}$$

We note however that the accuracy by which the entropy can be determined from simulations is an order of magnitude smaller than that of the free energy [13,35]. The *energy* difference ΔE_{BA} can be computed from the formula

$$\Delta E_{BA} = \Delta F_{BA} + T \Delta S_{BA} \tag{5.5}$$

using the values obtained for the free energy difference and the entropy difference by formulae (4.14) and (5.4). The error bars on the energy difference will be as large as those on the entropy difference.

6. Free energy difference between two non-equilibrium states A and B using a restraining potential

The formulae for the free energy difference between two states A and B as given in the previous sections have been cast in terms of the free energy of the total system, that is, including all degrees of freedom of the system. Since the system will always strive for a minimum of the free energy, states A and B can only be minimum free energy or equilibrium states of the system. In a number of cases one is interested in free energy differences between non-equilibrium states of a system, for example, states along a postulated reaction coordinate. In that case the relevant quantity is the free energy $F_d(d')$ of the system along the specified coordinate d, as a function of the value d'. The specified coordinate d can generally be written as a function of the coordinates **r** of the atoms

$$d \equiv d(\mathbf{r}) \tag{6.1}$$

The free energy $F_d(d')$ is defined by

$$F_d(d') = -k_B T \ln Z_d(d') \tag{6.2}$$

in terms of the partition function

$$Z_d(d') = [h^{3N-1} N!]^{-1} \iint \exp[-H(\mathbf{p},\mathbf{r})/k_B T]\, \delta(d-d')\, d\mathbf{p} d\mathbf{r} \tag{6.3}$$

where $\delta(d-d')$ is the Dirac delta function. If the value d' of the d-coordinate occurs frequently during a simulation of the system, that is, if

$$d' \approx d(F = \text{minimum}) \tag{6.4}$$

the value of $F_d(d')$ can be obtained from the probability

$$\pi(d')\ \alpha\ \exp[-F_d(d')/k_B T] \tag{6.5}$$

This formula cannot be used if d' corresponds to a non-equilibrium state, which is poorly sampled in an equilibrium simulation. In this case, an extra d-coordinate restraining potential energy term $V_d^*(d,d')$, that keeps the specified d-coordinate near the required value d', must be added to the potential energy (3.1).

In this section we will derive a thermodynamic integration formula for the free energy difference $F_d(d_B) - F_d(d_A)$ between two non-equilibrium states A and B of a system, that corresponds to d-coordinate values d_A and d_B.

In order to carry the system over from a state with $d' = d_A$ to one with $d' = d_B$, the extra d-coordinate restraining potential energy term is made a function of the coupling parameter λ, for example

$$V_d^*(d,\lambda) \equiv \frac{1}{2} K_d^*(\lambda)\, [d - d_0^*(\lambda)]^2 \tag{6.6}$$

where $K_d^*(\lambda)$ is the force constant and $d_0^*(\lambda)$ is the value to which the d-coordinate is restrained. The Hamiltonian including this restraining term becomes

$$H^*(\mathbf{p},\mathbf{r},\lambda) = H(\mathbf{p},\mathbf{r}) + V_d^*(d(\mathbf{r}),\lambda) \tag{6.7}$$

The symbol * is used to denote quantities that are computed using the force field (3.1) *plus* the extra restraining potential (6.6). The corresponding partition function becomes

$$Z_d^*(d',\lambda) = [h^{3N-1}N!]^{-1} \iint \exp[-H^*(\mathbf{p},\mathbf{r},\lambda)/k_BT]\,\delta(d-d')\,d\mathbf{p}d\mathbf{r} \tag{6.8}$$

from which the free energy in the presence of the d-coordinate restraining potential energy term can be obtained

$$F_d^*(d',\lambda) = -k_BT \ln Z_d^*(d',\lambda) \tag{6.9}$$

In order to obtain the free energy difference $F_d(d_B) - F_d(d_A)$ between two non-equilibrium states defined by d_A and d_B, the λ-dependence of $V_d^*(d,\lambda)$ must be chosen such that

$$d_0^*(\lambda_A) = d_A \tag{6.10}$$

$$d_0^*(\lambda_B) = d_B \tag{6.11}$$

and such that $K_0^*(\lambda_A)$, respectively $K_0^*(\lambda_B)$, are sufficiently large to keep the system near d_A, respectively near d_B. From a simulation including the restraining term $V_d^*(d,\lambda)$, the free energy difference

$$F_d^*(d_B,\lambda_B) - F_d^*(d_A,\lambda_A) = \int_{\lambda_A}^{\lambda_B} \frac{\partial F_d^*(d',\lambda)}{\partial\lambda}\,d\lambda \tag{6.12}$$

can be obtained using the formula

$$\frac{\partial F_d^*(d',\lambda)}{\partial\lambda} = \left\langle \frac{\partial V_d^*(d,\lambda)}{\partial\lambda} \right\rangle_{H^*(\mathbf{p},\mathbf{r},\lambda);d=d'} \tag{6.13}$$

The difference between $F_d(d')$, the free energy in the absence of the restraining potential, and $F_d^*(d',\lambda)$, the free energy in the presence of the restraining potential, is

$$F_d(d') - F_d^*(d',\lambda) = -k_BT \ln \frac{Z_d(d')}{Z_d^*(d',\lambda)}$$

$$= -k_BT \ln \left\{ \frac{\iint \exp[-H(\mathbf{p},\mathbf{r},)/k_BT]\, \delta(d-d')\, d\mathbf{p}d\mathbf{r}}{\iint \exp[-(H(\mathbf{p},\mathbf{r}) + V_d^*(d(\mathbf{r}),\lambda))/k_BT]\, \delta(d-d')\, d\mathbf{p}d\mathbf{r}} \right\}$$

$$= -k_BT \ln\{ < \exp[+V_d^*(d(\mathbf{r}),\lambda)/k_BT] >_{H(\mathbf{p},\mathbf{r})+V_d^*(d(\mathbf{r}),\lambda);d=d'} \}$$

$$= -V_d^*(d',\lambda) \tag{6.14}$$

So, the thermodynamic *integration formula* for the free energy difference between two *non-equilibrium states* A and B (λ_A and λ_B), defined by the values d_A and d_B of a specified coordinate d (**r**), which is based on an ensemble which is generated using a d-coordinate *restraining potential* energy term $V_d^*(d,\lambda)$ reads

$$F_d(d_B) - F_d(d_A) =$$

$$\int_{\lambda_A}^{\lambda_B} < \frac{\partial V_d^*(d(\mathbf{r}),\lambda)}{\partial \lambda} >_{H(\mathbf{p},\mathbf{r})+V_d^*(d(\mathbf{r}),\lambda);d=d'} d\lambda \; - V_d^*(d_B,\lambda_B) + V_d^*(d_A,\lambda_A) \tag{6.15}$$

In the ensemble average only contributions of configurations with $d(\mathbf{r}) = d'$ must be counted.

7. Thermodynamic cycles

The next step when using the thermodynamic integration technique to calculate relative free energies, or binding constants of donor-acceptor complexes, or solubilities is to formulate a so-called thermodynamic cycle. The basis on which the thermodynamic cycle approach rests is the fact that the (Helmholtz) free energy F is a thermodynamic state function. This means that as long as a system is changed in a reversible way the change in free energy ΔF will be independent of the path. Therefore, along a closed path or cycle one has $\Delta F = 0$. This result implies that there are two possibilities of obtaining ΔF for a specific process; one may calculate it directly using the techniques discussed above along a path corresponding to the process, or one may design a cycle of which the specific process is only a part and calculate the ΔF of the remaining part of the cycle.

The power of this *thermodynamic cycle technique* lies in the fact that on the computer also *non-chemical processes* such as the conversion of one type of atom into another type may be performed.

In order to visualize the method, we consider the relative binding of two inhibitors I_A and I_B to an enzyme E. The appropriate thermodynamic cycle for obtaining the relative binding constant is

$$\begin{array}{ccccc} & E + I_A & \xrightarrow{\text{1 (exp.)}} & (E:I_A) & \\ \text{3 (sim.)} & \downarrow & & \downarrow & \text{4 (sim.)} \\ & E + I_B & \xrightarrow[\text{2 (exp.)}]{} & (E:I_B) & \end{array} \qquad (7.1)$$

where the symbol : means complex formation. The relative binding constant equals

$$K_2/K_1 = e^{-[\Delta F_2 - \Delta F_1]/RT} \qquad (7.2)$$

where R denotes the gas constant. However, simulation of processes 1 and 2 is virtually impossible since it would involve the removal of many solvent molecules from the binding site of the inhibitor on the enzyme to be substituted by the inhibitor. But, since (7.1) is a cycle we have

$$\Delta F_2 - \Delta F_1 = \Delta F_4 - \Delta F_3 \qquad (7.3)$$

and, if the composition of inhibitor I_B is not too different from that of I_A, the desired result can be obtained by simulating the non-chemical processes 3 and 4.

8. Computation of free energy in practice

It will be clear that the applications of the thermodynamic cycle integration technique are manifold in chemistry. With an eye to future applications of the technique we list below a number of approximations and accuracy limiting factors which should be taken into consideration and discussed when performing free energy calculations for a specific system.

8.1. Use of generalized coordinates and momenta

Classical statistical mechanics is based on detailed classical dynamics of the atoms of the molecular system that is considered. The classical equations of motion are Newton's equations of motion (3.3-4), where the potential energy function $V(\mathbf{r}_1,\mathbf{r}_2,...,\mathbf{r}_N)$ has been assumed to be only a function of the Cartesian coordinates $\mathbf{r}_i$ of the N atoms ($i=1,2,...,N$). The kinetic energy is given by

$$K(\dot{\mathbf{r}}) = \sum_{i=1}^{N} \tfrac{1}{2} m_i \dot{\mathbf{r}}_i^2 \qquad (8.1.1)$$

where the velocities are denoted by

$$\dot{\mathbf{r}}_i \equiv \frac{d\mathbf{r}_i}{dt} \qquad i=1,2,...,N \qquad (8.1.2)$$

and $\mathbf{r}$ is a short-hand notation for the set $\{\mathbf{r}_1,\mathbf{r}_2,...,\mathbf{r}_N\}$. The form of Newton's equations of motion depends on the coordinate system which is used. The formulae (3.3-4) are only valid in a Cartesian coordinate system. But, classical mechanics can be formulated in terms of the Hamilton formalism, which is independent of the coordinate system that is used.

The Hamilton formulation of classical mechanics is the following [38,39]. *Generalized coordinates* $\mathbf{q}$ are denoted by

$$\mathbf{q} \equiv \{q_1,q_2,...,q_M\} \qquad (8.1.3)$$

for a system of M degrees of freedom. *Generalized momenta*

$$\mathbf{p} \equiv \{p_1,p_2,...,p_M\} \qquad (8.1.4)$$

conjugate to the generalized coordinates $\mathbf{q}$ are defined by

$$p_i = \frac{\partial[K(\dot{\mathbf{q}},\mathbf{q}) - V(\mathbf{q})]}{\partial \dot{q}_i} \qquad i=1,2,...,M \qquad (8.1.5)$$

where

$$\dot{q}_i \equiv \frac{dq_i}{dt} \qquad i=1,2,...,M \qquad (8.1.6)$$

In a Cartesian coordinate system the kinetic energy K is only a function of the velocities $\dot{\mathbf{r}}$, but in an arbitrary coordinate system it may depend on the generalized coordinates $\mathbf{q}$ as well as on the generalized velocities $\dot{\mathbf{q}}$. The Hamilton function or Hamiltonian is defined as

$$H(\mathbf{p},\mathbf{q}) = \sum_{i=1}^{M} p_i \dot{q}_i - K(\dot{\mathbf{q}},\mathbf{q}) + V(\mathbf{q})$$

$$= K(\mathbf{p},\mathbf{q}) + V(\mathbf{q}) \qquad (8.1.7)$$

The Hamilton equations of motion are

$$\frac{\partial H(\mathbf{p},\mathbf{q})}{\partial p_i} = \dot{q}_i \quad \text{and} \quad \frac{\partial H(\mathbf{p},\mathbf{q})}{\partial q_i} = -\dot{p}_i \qquad (8.1.8)$$

In the previous sections, the statistical mechanical formulae for the free energy and entropy have been cast in terms of the Hamilton formalism. This means that once the generalized coordinates $\mathbf{q}$ have been chosen, the generalized momenta $\mathbf{p}$ are defined through (8.1.5), and the Hamiltonian must be expressed as a function of $\mathbf{p}$ and $\mathbf{q}$ before the derivative with respect to the coupling parameter λ is taken.

How essential the requirement of expressing the Hamiltonian H as a function of $\mathbf{q}$ and the *conjugate* momenta $\mathbf{p}$ is, is illustrated by the following example. We consider a simple one-atom system, of which only the mass is taken to be dependent on the coupling parameter λ, see Formula (4.2)

$$m(\lambda) = [1-\lambda]\, m^A + \lambda\, m^B \qquad (8.1.9)$$

We choose the Cartesian coordinates $\mathbf{r}$ as generalized coordinates $\mathbf{q}$, so $\mathbf{q}=\mathbf{r}$, $\dot{\mathbf{q}}=\dot{\mathbf{r}}$ and

$$K(\dot{\mathbf{r}},\mathbf{r}) - V(\mathbf{r}) = \frac{1}{2}\, m(\lambda)\, \dot{\mathbf{r}}^2 - V(\mathbf{r}) \qquad (8.1.10)$$

The generalized momenta $\mathbf{p}_r$ conjugate to the generalized coordinates $\mathbf{r}$ are defined by (8.1.5)

$$\mathbf{p}_r = m(\lambda)\, \dot{\mathbf{r}} \qquad (8.1.11)$$

and the Hamiltonian expressed in the generalized coordinates $\mathbf{r}$ and the conjugate momenta $\mathbf{p}_r$ reads

$$H(\mathbf{p}_r,\mathbf{r},\lambda) = \frac{\mathbf{p}_r^2}{2m(\lambda)} + V(\mathbf{r}) \tag{8.1.12}$$

from which follows that

$$\frac{\partial H(\mathbf{p}_r,\mathbf{r},\lambda)}{\partial\lambda} = -\frac{1}{2}[m^B - m^A]\,\dot{\mathbf{r}}^2 \tag{8.1.13}$$

If we would have expressed the Hamiltonian in terms of the generalized coordinates $\mathbf{r}$ and (non-conjugate) velocities $\dot{\mathbf{r}}$

$$H(\dot{\mathbf{r}},\mathbf{r},\lambda) = \tfrac{1}{2}\, m(\lambda)\,\dot{\mathbf{r}}^2 + V(\mathbf{r}) \tag{8.1.14}$$

we would have obtained

$$\frac{\partial H(\dot{\mathbf{r}},\mathbf{r},\lambda)}{\partial\lambda} = +\frac{1}{2}[m^B - m^A]\,\dot{\mathbf{r}}^2 \tag{8.1.15}$$

which is clearly different from the correct Formula (8.1.13).

8.2. Choice of coordinate system

The use of a Cartesian coordinate system has some advantages. If the potential energy function V is conservative, that is, if $V(\mathbf{r})$ is not dependent on the velocities $\dot{\mathbf{r}}$, the integral over the momenta $\mathbf{p}_r$ in the partition function (2.2)

$$Z = [h^{3N}N!]^{-1}\iint \exp[-H(\mathbf{p}_r,\mathbf{r})/k_BT]\,d\mathbf{p}_r d\mathbf{r} \tag{8.2.1}$$

reduces to a product of factors which are independent of the coordinates $\mathbf{r}$. Using

$$H(\mathbf{p}_r,\mathbf{r}) = \sum_{i=1}^{N}\frac{\mathbf{p}_{ri}^2}{2m_i} + V(\mathbf{r}) \tag{8.2.2}$$

we find

$$Z = [h^{3N}N!]^{-1} \int \exp[-\sum_{i=1}^{N} \mathbf{p}_{ri}^2/(2m_i k_B T)]\, d\mathbf{p}_r \int \exp[-V(\mathbf{r})/k_B T]\, d\mathbf{r}$$

$$= [2\pi h^{-2} k_B T]^{3N/2} [N!]^{-1} \left[\prod_{i=1}^{N} m_i\right]^{3/2} \int \exp[-V(\mathbf{r})/k_B T]\, d\mathbf{r} \qquad (8.2.3)$$

where the integral over configuration space is called the *configurational integral.*

If instead of Cartesian coordinates arbitrary coordinates $\mathbf{q}$ are used, the expression for the configurational integral becomes more complicated. We assume that the 3N Cartesian coordinates

$$\underline{x} \equiv (x_1, x_2, \ldots, x_{3N})^\tau \qquad (8.2.4)$$

of the N atoms can be expressed as a (matrix) function $\underline{x}(\underline{q})$ of the 3N generalized coordinates

$$\underline{q} = (q_1, q_2, \ldots, q_{3N})^\tau \qquad (8.2.5)$$

where we have employed matrix notation. Transformation of the kinetic energy from Cartesian coordinates to $\underline{q}$-coordinates leads to

$$K(\underline{\dot{q}}, \underline{q}) = \frac{1}{2} \sum_{k=1}^{3N} m_k \dot{x}_k^2$$

$$= \frac{1}{2} \sum_{i=1}^{3N} \sum_{j=1}^{3N} \sum_{k=1}^{3N} \dot{q}_i \frac{\partial x_k}{\partial q_i} m_k \frac{\partial x_k}{\partial q_j} \dot{q}_j$$

$$= \frac{1}{2} \underline{\dot{q}}^\tau \underline{G} \underline{\dot{q}} \qquad (8.2.6)$$

where the *mass-metric tensor* corresponding to the coordinate transformation

$$\underline{x} = \underline{x}(\underline{q}) \qquad (8.2.7)$$

is defined as

$$G_{ij} = \sum_{k=1}^{3N} m_k \frac{\partial x_k}{\partial q_i} \frac{\partial x_k}{\partial q_j} \qquad i,j = 1,2,...,3N \tag{8.2.8}$$

The momenta $\underline{p}$ conjugate to the coordinates $\underline{q}$ become

$$\underline{p} = \frac{\partial K(\dot{\underline{q}},\underline{q})}{\partial \dot{\underline{q}}} = \underline{G}\,\dot{\underline{q}} \tag{8.2.9}$$

and the Hamiltonian expressed in $\underline{p}$ and $\underline{q}$ is

$$H(\underline{p},\underline{q}) = \frac{1}{2}\underline{p}^T \underline{G}^{-1} \underline{p} + V(\underline{q}) \tag{8.2.10}$$

The partition function is given by

$$Z = [h^{3N}N!]^{-1} \iint \exp[-H(\underline{p},\underline{q})/k_BT]\, d\underline{p}d\underline{q} \tag{8.2.11}$$

for any choice of *canonical variables* $\underline{p}$ and $\underline{q}$ in full phase space, since the volume element $d\underline{p}d\underline{q}$ is invariant for canonical transformations [39]. Using (8.2.10) the integration over the momenta $\underline{p}$ can be carried out

$$Z = [h^{3N}N!]^{-1} \iint \exp[-\underline{p}^T\underline{G}^{-1}\underline{p}/2k_BT] \exp[-V(\underline{q})/k_BT]\, d\underline{p}d\underline{q}$$

$$= [2\pi h^{-2}k_BT]^{3N/2} [N!]^{-1} \int |\underline{G}(\underline{q})|^{1/2} \exp[-V(\underline{q})/k_BT]\, d\underline{q} \tag{8.2.12}$$

where the determinant of the mass-metric tensor $\underline{G}$ is denoted by $|\underline{G}|$. The type of coordinate transformation $\underline{x}(\underline{q})$ will determine whether the mass-metric tensor depends on the generalized coordinates $\underline{q}$. If this is the case, the square root of its determinant enters as a weight function in the configurational integral.

8.3. Application of constraints

In computer simulation of molecular systems, it is often computationally efficient to reduce the number of degrees of freedom. This can be achieved by eliminating from the system the hard degrees of freedom, viz., those corresponding to high-frequency normal modes. Constraining the bond lengths of a molecule to fixed values generally increases the computational efficiency by a factor of 2 to 3 by allowing for a larger integration time step Δt in the simulation.

When constraints are applied in a computer simulation, or in any calculation of statistical mechanical averages, a subtle problem arises. A constrained system is described in generalized coordinates and momenta by a Hamiltonian that does *not* include those quantities of the constrained degrees of freedom. This is not equivalent to a Hamiltonian that is obtained as a limiting case of a Hamiltonian of full dimensionality, in which the corresponding degrees of freedom are considered as harmonic oscillators in the limit of infinite force constants. In the latter case, the momenta conjugate to the hard degrees of freedom do not vanish, as they do in the constrained case. For systems with internal degrees of freedom, where constrained degrees of freedom are not orthogonal to non-constrained ones, the consequence is that in constrained dynamics different regions of configuration space may get different weight factors depending on the Jacobian of the transformation from Cartesian to generalized coordinates. For a discussion of the problem we refer to Refs. 40-44. The metric tensor problem does not occur when rigid bodies without internal degrees of freedom are simulated.

In this subsection we derive an expression for the free energy difference ΔF_{BA} for a molecular model which involves both non-constrained and rigid (constrained) internal degrees of freedom.

As in the previous subsection, we consider a transformation from 3N Cartesian coordinates $\underline{x}$ to $M=3N$ generalized coordinates $\underline{q}$, which are divided into N_β hard variables $\underline{q}^\beta$, e.g., bond lengths of a molecule, and N_α soft variables $\underline{q}^\alpha$, e.g., the remaining degrees of freedom of the molecule. Clearly one has $N_\alpha+N_\beta=M=3N$. In a *rigid model* of the molecule the hard variables $\underline{q}^\beta$ are rigidly constrained to fixed values $\underline{q}^\beta_0$. This means that in the full 6N dimensional Cartesian phase space the Cartesian coordinates $\underline{x}$ and momenta $\underline{p}_x$ are constrained to a hypersurface for which $\underline{q}^\beta=\underline{q}^\beta_0$ and $\underline{p}^\beta=\underline{0}$ (momenta along constrained degrees of freedom equal zero). When performing the transformation from $\underline{x}$ to $\underline{q}$ the hard variables $\underline{q}^\beta$ and their conjugate momenta $\underline{p}^\beta$ are removed from the Hamiltonian

$$H^\alpha(\underline{p}^\alpha,\underline{q}^\alpha) = \tfrac{1}{2}(\underline{p}^\alpha)^T(\underline{G}^\alpha)^{-1}\,\underline{p}^\alpha+V(\underline{q}^\alpha) \qquad (8.3.1)$$

where the mass-metric tensor is now limited to the soft variables $\underline{q}^\alpha$

$$G^\alpha_{ij} = \sum_{k=1}^{3N} m_k \frac{\partial x_k}{\partial q^\alpha_i}\frac{\partial x_k}{\partial q^\alpha_j} \qquad i,j=1,2,\ldots,N_\alpha \qquad (8.3.2)$$

and the momenta $\underline{p}^\alpha$ conjugate to $\underline{q}^\alpha$ are

$$\underline{p}^\alpha = \underline{G}^\alpha\,\dot{\underline{q}}^\alpha \qquad (8.3.3)$$

like in the full phase space treatment in the previous subsection.

The next step is to make the rigid model Hamiltonian $H^\alpha(\underline{p}^\alpha,\underline{q}^\alpha)$ dependent on the coupling parameter λ. If the hard variables $\underline{q}^\beta$ are chosen to be λ-dependent, e.g. if λ controls the change of the length of a rigid bond in the molecule, the transformation from Cartesian coordinates $\underline{x}$ to the generalized ones $\underline{q}^\alpha$ will also become dependent on λ:

$$\underline{x} = \underline{x}(\underline{q}^\alpha,\lambda) \tag{8.3.4}$$

and so will the mass-metric tensor $\underline{G}^\alpha(\lambda)$ corresponding to the transformation. The λ-dependent Hamiltonian reads

$$H^\alpha(\underline{p}^\alpha,\underline{q}^\alpha,\lambda) = \frac{1}{2}(\underline{p}^\alpha)^T\,(\underline{G}^\alpha(\underline{q}^\alpha,\lambda))^{-1}\,\underline{p}^\alpha + V(\underline{q}^\alpha,\lambda) \tag{8.3.5}$$

which yields for the rigid model partition function after integration over the conjugate soft momenta $\underline{p}^\alpha$, like in the previous subsection,

$$Z^\alpha(\lambda) = [\,2\pi h^{-2} k_B T\,]^{N_\alpha/2} \int |\underline{G}^\alpha(\underline{q}^\alpha,\lambda)|^{1/2} \exp[-V(\underline{q}^\alpha,\lambda)/k_BT]\, d\underline{q}^\alpha \tag{8.3.6}$$

where we have omitted the factor $[N_\alpha!]^{-1}$, since the atoms in the molecule are distinguishable. Taking the derivative of the rigid model free energy

$$F^\alpha(\lambda) = -k_BT \ln Z^\alpha(\lambda) \tag{8.3.7}$$

with respect to λ we find

$$\left[\frac{\partial F^\alpha(\lambda)}{\partial\lambda}\right]_T =$$

$$+\left\{\frac{\int \frac{\partial V(\underline{q}^\alpha,\lambda)}{\partial\lambda}\, |\underline{G}^\alpha(\underline{q}^\alpha,\lambda)|^{1/2} \exp[-V(\underline{q}^\alpha,\lambda)/k_BT]\, d\underline{q}^\alpha}{\int |\underline{G}^\alpha(\underline{q}^\alpha,\lambda)|^{1/2} \exp[-V(\underline{q}^\alpha,\lambda)/k_BT]\, d\underline{q}^\alpha}\right\}$$

$$-k_BT\left\{\frac{\int \frac{\partial |\underline{G}^\alpha(\underline{q}^\alpha,\lambda)|^{1/2}}{\partial\lambda} \exp[-V(\underline{q}^\alpha,\lambda)/k_BT]\, d\underline{q}^\alpha}{\int |\underline{G}^\alpha(\underline{q}^\alpha,\lambda)|^{1/2} \exp[-V(\underline{q}^\alpha,\lambda)/k_BT]\, d\underline{q}^\alpha}\right\} \tag{8.3.8}$$

The first term is the usual ensemble average of $\partial V(\lambda)/\partial\lambda$, but the probability distribution function contains an extra configuration and λ-dependent weight factor $|\underline{G}^{\alpha}(\underline{q}^{\alpha},\lambda)|^{1/2}$. The second term is an ensemble average (including the same weight factor) of

$$\frac{\partial|\underline{G}^{\alpha}(\underline{q}^{\alpha},\lambda)|^{1/2}}{\partial\lambda} \Big/ |\underline{G}^{\alpha}(\underline{q}^{\alpha},\lambda)|^{1/2} = \frac{1}{2}\,\frac{\partial[\ln|\underline{G}^{\alpha}(\underline{q}^{\alpha},\lambda)|]}{\partial\lambda} \tag{8.3.9}$$

Metric tensor corrections of the type encountered in the first term of (8.3.8) turn out to be non-negligible in the case of application of bond-angle constraints [42-44]. However, when only bond lengths are constrained, the metric tensor effect is negligible, since the metric tensor weight factor depends only on the bond-angle degrees of freedom which are generally restricted to a narrow range of bond-angle values. The configurational dependence of $|\underline{G}^{\alpha}(\underline{q}^{\alpha},\lambda)|$ can be safely ignored in this case. The factor $|\underline{G}^{\alpha}(\lambda)|^{1/2}$ can be removed from the integrals in the numerator and the denominator, and will subsequently vanish from the first term in (8.3.8). Using the same approximation of the independence of $|\underline{G}^{\alpha}(\underline{q}^{\alpha},\lambda)|$ on the configuration $\underline{q}^{\alpha}$, the second term in (8.3.8) reduces to

$$-\frac{1}{2}k_BT\,\frac{\partial|\underline{G}^{\alpha}(\lambda)|}{\partial\lambda} \tag{8.3.10}$$

which can directly be integrated over λ in order to obtain its contribution to ΔF_{BA}. When calculating the free energy contributions from different parts of a thermodynamic cycle, terms of the type (8.3.10) may effectively cancel.

We conclude that, when constraining bond lengths in the dynamics of a molecular system, metric tensor corrections are not required when calculating relative free energy differences. When absolute free energies are to be obtained, the extra term (8.3.10) has to be evaluated. On the other hand, when constraining bond angles in addition to bond lengths metric tensor corrections to the free energy will be significant.

8.4. Computation of the contribution of constraint forces to the free energy

Various methods are available for applying constraints to a molecular system [1,2,40]. A simple method is the procedure called SHAKE [40,45]. Its essential feature is that after each MD integration time step Δt the constraints are satisfied by adding displacement vectors to the (Cartesian) position vectors of the atoms that result from a non-constrained time step. The displacement vectors are determined such that the constraints are satisfied at the final positions. We will denote the application of SHAKE by

$$\text{SHAKE}\ (\mathbf{r}(t_n),\ \mathbf{r}'(t_{n+1}),\ \mathbf{r}(t_{n+1})) \tag{8.4.1}$$

As before, the set $\{\mathbf{r}_1,\mathbf{r}_2,...,\mathbf{r}_N\}$ of position vectors of the N atoms is denoted by $\mathbf{r}$. Formula (8.4.1) means that the (*non-constrained*) atomic positions $\mathbf{r}'(t_{n+1})$ that result from the non-constrained time step from time t_n to time $t_{n+1} \equiv t_n + \Delta t$ will be reset to give the *constrained* positions $\mathbf{r}(t_{n+1})$. The direction of the displacement vectors

$$\delta\mathbf{r} = \mathbf{r}(t_{n+1}) - \mathbf{r}'(t_{n+1}) \tag{8.4.2}$$

is determined by the *reference* positions $\mathbf{r}(t_n)$; that is, for each individual constraint the displacement of the pair of atoms involved is parallel to the vector from one atom to the other in the reference configuration.

The *constraint forces* $\mathbf{G}(t_n)$ can be obtained from the difference $\delta\mathbf{r}$ between the constrained positions (resulting from a constrained time step Δt) and the non-constrained positions (resulting from a non-constrained time step Δt) at time t_{n+1}

$$\mathbf{G}(t_n) = m\ \delta\mathbf{r} / (\Delta t)^2 \tag{8.4.3}$$

where the index denoting atoms has been omitted as in the remainder of this subsection. The work done by the constraint forces $\mathbf{G}(t_n)$ during the time step Δt from t_n to t_{n+1} is

$$W = \mathbf{G}(t_n) \cdot [\mathbf{r}(t_{n+1}) - \mathbf{r}(t_n)] \tag{8.4.4}$$

In the thermodynamic integration formalism the derivative of this work with respect to the coupling parameter λ is required. Since the work done by the constraint forces is not available in analytical form, a numerical expression for the derivative is used

$$\left[\frac{\partial W(\lambda)}{\partial \lambda}\right]_{\lambda=\lambda_i} = \frac{W(\lambda_{i+1}) - W(\lambda_{i-1})}{\lambda_{i+1} - \lambda_{i-1}} \tag{8.4.5}$$

where consecutive values of λ are denoted by λ_i ($i = 1,2,...$). Using the dependence of the coupling parameter λ on the time t, $\lambda(t)$, one has $\lambda_n \equiv \lambda(t_n)$. When the constraints, e.g., the bond lengths $b_0(\lambda)$, depend on λ, the positions $\mathbf{r}(t,\lambda)$ resulting from application of the procedure SHAKE will also depend on the value of λ, and so will the constraint forces $\mathbf{G}(t,\lambda)$ and the work $W(\lambda)$ done by these forces.

In order to obtain the work $W(\lambda)$ at the value λ_{n+1}, the integration step and the application of SHAKE has to be repeated:

- calculate the force $\mathbf{F}(t_n,\lambda_{n+1})$ from the potential energy function $V(\mathbf{r}(t_n),\lambda_{n+1})$ and compute non-constrained positions $\mathbf{r}'(t_{n+1},\lambda_{n+1})$ using this force;
- perform

$$\text{SHAKE}\,(\,\mathbf{r}(t_n,\lambda_n),\ \mathbf{r}'(t_{n+1},\lambda_{n+1}),\ \mathbf{r}(t_{n+1},\lambda_{n+1})\,) \tag{8.4.6}$$

- compute the constraint forces from the difference between constrained and non-constrained positions (8.4.2)

$$\mathbf{G}(t_n,\lambda_{n+1}) = [\,\mathbf{r}(t_{n+1},\lambda_{n+1}) - \mathbf{r}'(t_{n+1},\lambda_{n+1})\,] \tag{8.4.7}$$

- compute the work done by the constraint forces

$$W(\lambda_{n+1}) = \mathbf{G}(t_n,\lambda_{n+1})\cdot[\,\mathbf{r}(t_{n+1},\lambda_{n+1}) - \mathbf{r}(t_n,\lambda_n)\,] \tag{8.4.8}$$

In order to obtain the work $W(\lambda)$ at the value λ_{n-1}, this procedure is to be repeated with λ_{n-1} instead of λ_{n+1}:

- calculate the force $\mathbf{F}(t_n,\lambda_{n-1})$ from the potential energy function $V(\mathbf{r}(t_n),\lambda_{n-1})$ and compute non-constrained positions $\mathbf{r}'(t_{n+1},\lambda_{n-1})$ using this force;
- perform

$$\text{SHAKE}\,(\,\mathbf{r}(t_n,\lambda_n),\ \mathbf{r}'(t_{n+1},\lambda_{n-1}),\ \mathbf{r}(t_{n+1},\lambda_{n-1})\,) \tag{8.4.9}$$

- compute the constraint forces from the difference between constrained and non-constrained positions (8.4.2)

$$\mathbf{G}(t_n,\lambda_{n-1}) = [\,\mathbf{r}(t_{n+1},\lambda_{n-1}) - \mathbf{r}'(t_{n+1},\lambda_{n-1})\,] \tag{8.4.10}$$

- compute the work done by the constraint forces

$$W(\lambda_{n-1}) = \mathbf{G}(t_n,\lambda_{n-1})\cdot[\,\mathbf{r}(t_{n+1},\lambda_{n-1}) - \mathbf{r}(t_n,\lambda_n)\,] \tag{8.4.11}$$

The contribution of the constraint forces during the time step from t_n to $t_n + \Delta t$ to the free energy difference ΔF_{BA} becomes

$$\Delta F^c_n(\Delta t) = \tfrac{1}{2}\,[\,W(\lambda_{n+1}) - W(\lambda_{n-1})\,] \tag{8.4.12}$$

Application of this procedure requires two extra force evaluations (at λ_{n-1} and at λ_{n+1}) and two extra calls of the procedure SHAKE. The extra force evaluations will not be computer time consuming, since these only involve perturbed atoms, that is, the terms in the potential energy function that depend on the coupling parameter λ. The calls of SHAKE will involve all atoms that are connected by an uninterrupted chain of constraints to the perturbed (λ-dependent) atoms. However, in most practical applications, one is not interested in the contribution of the constraint forces to the free energy, since these are likely to cancel when relative free energy differences are computed.

8.5. Creation or annihilation of atoms

When the states A and B between which the free energy difference ΔF_{BA} is to be calculated, differ in the number of atoms, atoms are to be created or annihilated. In that case, one may *not* choose m^A or m^B in (4.2) equal to zero, since Newton's equations of motion (3.3) contain the factor m^{-1}. One should just slowly reduce the interaction $V(\lambda)$ of the atom that is to be annihilated with the rest of the system to zero. By decoupling it from the rest of the system as a function of λ, the atom will gradually become a ghost atom.

In case of application of constraints, such as bond-length constraints, a complication arises. A constraint is either present or not present, its presence cannot be made a smooth function of the coupling parameter λ. This means that when an atom is involved in a constraint it cannot be completely decoupled from the system, since even when the interaction $V=0$, the constraint forces remain present. The constraint forces may be reduced by a reduction of the constrained bond length or of the atomic mass. However, a too large reduction will cause problems. If $b_0(\lambda)=0$, two atoms will coincide, which will render bond angles involving the two atoms undefined. When the atomic mass is made smaller, the atomic velocity gets larger, since the equipartition theorem states that

$$\langle mv^2 \rangle \approx k_B T \qquad (8.5.1)$$

Larger velocities will require smaller integration time steps Δt, so the atomic mass should not be reduced beyond the smallest mass present in the system.

8.6. Quantum corrections

In this paper we have used a classical treatment for the calculation of free energy and entropy. This is incorrect for modes with frequencies larger than $k_B T/h$. In molecular systems, such modes include vibrations involving covalent bonds, bond angles and torsional angles. The inclusion of quantum corrections in the computation of free energy and entropy has been discussed in Ref.13. Only vibrations with wavenumbers higher than 500 cm^{-1} involve appreciable corrections to the free energy. The quantum correction for the energy (mainly zero-point energy) and that for the entropy tend to cancel each other [13]. When the high-frequency range of the spectral density has the same form in state A as in state B, the quantum corrections will cancel when ΔF_{BA} is computed. In the unlikely case that the quantum correction to ΔF_{BA} is non-negligible, the contributions from different parts of a thermodynamic cycle may be comparable and so effectively cancel when the relative free energy difference between two processes $A \rightarrow B$ and $A' \rightarrow B'$ is determined.

We conclude that although quantum corrections are important when absolute values for the free energy or entropy are required, they can safely be disregarded in most practical applications.

8.7. Derivative of the Hamiltonian with respect to λ

The thermodynamic integration formula (4.14)

$$\Delta F_{BA} = \int_{\lambda_A}^{\lambda_B} \left< \frac{\partial H(\mathbf{p},\mathbf{q},\lambda)}{\partial \lambda} \right>_\lambda d\lambda \tag{8.7.1}$$

requires the evaluation of an integral over the ensemble average of the derivative of the Hamiltonian with respect to λ. When applying MD to generate the ensemble, the coupling parameter λ is made a function of time

$$\lambda = \lambda(t) \tag{8.7.2}$$

and one obtains

$$\Delta F_{BA} = \sum_{n=1}^{N_{MD}} \left[\frac{\partial H(\mathbf{p},\mathbf{q},\lambda)}{\partial \lambda} \right]_{\lambda=\lambda_{n-1}} [\lambda_n - \lambda_{n-1}] \tag{8.7.3}$$

where N_{MD} is the number of MD time steps in the simulation and $\lambda_n = \lambda(t_n)$.

One may evaluate (8.7.3) in different ways. When the Hamiltonian is given as a function of λ in analytical form (4.1-8), the *analytical derivative* $\partial H(\mathbf{p},\mathbf{q},\lambda)/\partial\lambda$ can be used. An alternative is to use the *numerical derivative*

$$\left[\frac{\partial H(\mathbf{p},\mathbf{q},\lambda)}{\partial \lambda} \right]_{\lambda=\lambda_n} \approx \frac{H(\mathbf{p}(\lambda_n),\mathbf{q}(\lambda_n),\lambda_{n+1}) - H(\mathbf{p}(\lambda_n),\mathbf{q}(\lambda_n),\lambda_{n-1})}{\lambda_{n+1} - \lambda_{n-1}} \tag{8.7.4}$$

where the Hamiltonian $H(\mathbf{p},\mathbf{q},\lambda)$ must be evaluated using $\lambda = \lambda_{n-1}$ and $\lambda = \lambda_{n+1}$, while using the positions $\mathbf{q}(\lambda_n)$ and momenta at $\mathbf{p}(\lambda_n)$ at $\lambda = \lambda_n$. This leads to the formula

$$\Delta F_{BA} = \sum_{n=1}^{N_{MD}-1} \tfrac{1}{2}[\,H(\mathbf{p}(\lambda_n),\mathbf{q}(\lambda_n),\,\lambda_{n+1}) - H(\mathbf{p}(\lambda_n),\mathbf{q}(\lambda_n),\lambda_{n-1})\,] \tag{8.7.5}$$

In [45] the equation

$$\Delta F_{BA} = \sum_{n=1}^{N_{MD}} [\,H(\mathbf{p}(\lambda_n),\mathbf{q}(\lambda_n),\lambda_{n+1}) - H(\mathbf{p}(\lambda_n),\mathbf{q}(\lambda_n,\lambda_n)\,] \tag{8.7.6}$$

is used, which corresponds to taking the right-derivative of the Hamiltonian with respect to λ, instead of using the time-reversal invariant formula (8.7.5).

8.8. Thermodynamic integration versus thermodynamic perturbation

When applying the thermodynamic perturbation formula (4.12) a series of discrete MD simulations is performed at a series of λ-values $\lambda_1,\ldots,\lambda_n$. Using Formula (4.12) the free energy $F(\lambda_i \pm \Delta\lambda)$ in the neighbourhood of each value λ_i is determined and the n local curves $F(\lambda)$ for $\lambda_i - \Delta\lambda < \lambda < \lambda_i + \Delta\lambda$ are chained together to one curve for $F(\lambda)$ by fitting the overlapping parts $(F(\lambda_i + \Delta\lambda) \approx F(\lambda_{i+1} - \Delta\lambda))$. Therefore, this technique is also called the *window* $(\lambda \pm \Delta\lambda)$ or *discrete overlapping perturbations* technique. When applying the thermodynamic integration Formula (4.14), the coupling parameter λ is made a function of time, and so changes continuously and slowly during the MD simulation. Therefore, this technique is also called *slow-growth* or *continuous-coupling* technique. It has been demonstrated that application of the thermodynamic integration technique (4.14) is considerably more efficient than the overlapping perturbations one [7,29].

8.9. Choice of λ-dependence of the Hamiltonian $H(\lambda)$

The free energy F is a thermodynamic state function, so the free energy difference $F(\lambda_B) - F(\lambda_A)$ is independent of the path along which the system changes from state A to state B, as long as the path is traversed in a reversible way. The path is determined by the choice of the λ-dependence of $H(\lambda)$, like in (4.1-8). The rate of change of λ during the simulation is determined by the choice of the dependence of the coupling parameter $\lambda(t)$ on time t. Although one may choose any path, choosing a path with high-energy barriers between state A and state B will be very inefficient, since the changing of λ with time will have to be very slow in order to keep the change reversible when mounting the energy barrier. The optimum path is the one over which the change in free energy is constant over the whole path, that is, in which $F(\lambda(t))$ is a linear function of time. When creating an atom (state A = nothing, state B = atom) or growing a heavy atom (C, O, etc.) out of a hydrogen atom, one generally chooses $\lambda(t) \alpha t^5$ or t^6 in combination with the parameterization (4.8) for $V(r_{ij},\lambda)$; the growth must proceed very slowly in the early stages of atom creation, because of the steep r^{-12} repulsion of the growing atom by its neighbors [29,30]. If one would like to use a linear dependence of $\lambda(t)$ on time, another parameterization of $V(r_{ij},\lambda)$ should be used [47]. When growing Na^+ out of Ne, that is, when creating a charge, the choice $\lambda(t) \alpha t^{1/2}$ will lead to a linear behavior of $F(\lambda(t))$ [32,33]. It is advisable to carefully consider the choice of the λ-dependence of $H(\lambda)$ and of the time-dependence of $\lambda(t)$ with an eye to the physicochemical properties of the change from state A to state B, in order to obtain an efficient integration path.

8.10. Which terms in the interaction are taken to contribute to ΔF

When applying formula (4.12) or (4.14) all terms (4.2-8) in the Hamiltonian (4.1) which depend on the coupling parameter λ will contribute to the change in free energy ΔF_{BA}. However, in some applications [48,49] only the contribution of the nonbonded interaction term (4.8) to ΔF_{BA} is evaluated. This implies the assumption that the contribution of the intramolecular energy terms (4.4-7) to $\Delta F_{BA}$is identical for processes 3 and 4 in scheme (7.1), so that they cancel when $\Delta F_4 - \Delta F_3$ is evaluated. This assumption has not yet been verified. In view of the coupling between torsional degrees of freedom and nonbonded interaction ones, it would not be surprising when the assumption only holds for changes from state A to state B involving relatively rigid torsional angles.

8.11. Adequate sampling, or the relaxation time of the environment

The crucial factor in obtaining reliable free energy estimates from MD simulations lies in the adequacy of the sampling of phase or configuration space of the system. The energy of the system is obtained as an *average* over the ensemble of configurations, whereas the free energy is an *integral* over the phase or configuration space that is accessible to the system. This means that the reliability of the obtained free energy estimate strongly depends on the extent of the sampling of configuration space, especially when entropic effects do contribute to the free energy. Partial information on the adequacy of the sampling can be obtained by integrating in (4.14) not only forwards from state A to state B, but also backwards from state B to state A. If the change from state A to B to A is performed reversibility, the hysteresis will be zero: $\Delta F_{BA} + \Delta F_{AB} = 0$. Yet, this is only a necessary but not a sufficient condition for obtaining an accurate estimate of ΔF_{BA}. If the time period τ_{MD} of the MD simulation over which the change from state A to state B is performed is much longer than the relaxation time τ_{system} of the system ($\tau_{system} << \tau_{MD}$), the change can be carried out reversibly, and the hysteresis will be zero: $\Delta F_{BA} + \Delta F_{AB} \approx 0$. If $\tau_{system} \approx \tau_{MD}$, the process is not reversibly carried out, which will show up in significant hysteresis: $\Delta F_{BA} + \Delta F_{AB} \neq 0$. But, if $\tau_{system} >> \tau_{MD}$, the hysteresis $\Delta F_{BA} + \Delta F_{AB} \approx 0$, since the change is proceeding so fast that the system cannot adapt itself to it at all. In this case an incorrect free energy estimate is obtained despite the fact that $\Delta F_{BA} + \Delta F_{AB} \approx 0$. The system which is subject to a change in process 3 of scheme (7.1) may have quite a different relaxation time τ_3 compared to the relaxation time τ_4 of the system subject to a change in process 4. When in process 3 a part of the inhibitor molecule that sticks out into (aqueous) solution is changed, the relaxation time τ_3 is that of the solvent (water) molecule motion. The rotational correlation time of a water molecule is about 2 ps and the dielectric relaxation time of water is about 8 ps. This means that with simulations covering $\tau_{MD} > 20$ ps reasonably accurate estimates of ΔF_{BA} for process 3 can be obtained. However, in process 4 the changing part of the inhibitor molecule may be in contact with the enzyme, which may have a relaxation time beyond the 1-10 ps range, thus requiring a much longer MD simulation in order to obtain a reliable estimate of ΔF_{BA} of process 4. An example of this effect is given in Ref. 30. For an extensive discussion of criteria for sufficient sampling we refer to Ref. 32. We would recommend to repeat the MD simulation leading to an estimate of ΔF using different starting configurations of the system in order to obtain an impression of the accuracy of sampling. Moreover, one may apply different parameterizations of the Hamiltonian $H(\lambda)$ in order to trace a possible path-dependence of the obtained estimate of ΔF.

8.12. Effect of treatment of long-range interactions on ΔF

The Coulomb interaction in (3.1) and (4.8) is inversely proportional to the distance r_{ij} between atoms i and j. Due to this $1/r_{ij}$ distance dependence it is long-ranged. The size of the Coulomb interaction at a typical cut-off distance of 0.8-1.0 nm is even for partial atomic charges of size 0.1-0.4 *e* non-negligible. This would mean that nonbonded Coulomb interactions may only be neglected (cut off) beyond long distances, like 2.0-3.0 nm. The long-range character of the Coulomb interaction will strongly influence the estimate of ΔF_{BA} for processes ($A \rightarrow B$) in which a full atomic charge is created or annihilated [37]. This effect has been analyzed in Refs. 32 and 33. When changing a Ne atom (state A) dissolved in a box with water molecules into a Na^+ ion (state B), $\Delta F_{BA} = -421$ kJ/mol (actually ΔG_{BA}) was found when using a nonbonded interaction cut-off radius $R_c = 0.9$ nm, and $\Delta F_{BA} = -461$ kJ/mol when using $R_c = 1.2$ nm. So, an extension of the cut-off radius from 0.9 to 1.2 nm over only 0.3 nm yields a contribution of -40 kJ/mol to the free energy of solvation of the Na^+

ion. This effect can also be analyzed using the Born [50] formula for the free energy of ion hydration

$$\Delta F = -\frac{Z^2 e^2}{8\pi\epsilon_0 R} \frac{\epsilon_r - 1}{\epsilon_r} \tag{8.12.1}$$

which is derived by calculation of the reversible work required to charge (to a charge of Ze) a conducting hard sphere of radius R, embedded in a fluid at constant temperature and volume, treated as a polarizable dielectric with relative dielectric constant ϵ_r. Using $Z=1$, $\epsilon_r=80$ and the values $R=0.9$ nm and $R=1.2$ nm, the contribution of the hydration shell between 0.9 nm and 1.2 nm becomes -19 kJ/mol. Although the effect predicted by the simple continuum Born model is about half as large as that obtained from the MD simulations treating the aqueous solution in atomic detail, its absolute size is still of the order of 9 kT. One might argue that when a thermodynamic cycle is considered, the long-range Coulomb effects in processes 3 and 4 may be identical, and so cancel each other when $\Delta F_4 - \Delta F_3$ is calculated. However, this assumption strongly depends on the dielectric behavior of the inhibitor I in water (process 3) versus the inhibitor-enzyme (I:E) complex (process 4); the enzyme may well have quite different dielectric behavior than the water molecules that are surrounding the inhibitor in solution. In view of the large energies involved, we think that free energy calculations involving the creation or annihilation of atomic charges are rather model-dependent as long as the long-range Coulomb interaction is not properly taken into account. Fortunately, a number of methods are available to take into account long-range Coulomb interactions in MD simulations [51]. In case only neutral groups of atoms are substituted by neutral ones, the Coulomb interaction may be treated as a dipole-dipole interaction which has a $1/r_{ij}^3$ distance dependence, and so reduces the need to use a large cut-off radius considerably.

8.13. Appropriate treatment of boundary conditions

When deriving formulae (4.12 and 4.14) for ΔF_{BA} it has been assumed that the temperature T and the volume V of the molecular system are constant. As a consequence of this, the MD simulation must be carried out at constant $T=T_0$ and constant $V=V_0$, e.g., by coupling the system to a heat bath of the desired temperature T_0 [52]. Corresponding formulae for the Gibbs free energy difference ΔG_{BA} for a simulation at constant temperature $T=T_0$ and constant pressure $P=P_0$ can be found in Ref. 34. When simulating at constant volume V, the actual pressure of the system should be determined in order to allow for a comparison of the ΔF_{calc} calculated at constant (T,V) to the ΔG_{exp} generally measured at constant (T,P).

The classical way to minimize edge effects in simulations is to apply *periodic boundary conditions*. The atoms of the system are put into a cubic (or any periodically space filling shaped) box, which is surrounded by 26 identical translated images of itself. When calculating the forces on an atom in the central box, all interactions with atoms in the central box or images in the surrounding boxes that lie within the spherical cut-off, are taken into account. Thus in fact a crystal is simulated. For macromolecules application of periodic boundary conditions is far too expensive, due to the large amount of solvent (water) molecules that is needed to fill the box. In that case the number of atoms in the simulation can be limited by simulating only part of the molecule. Edge effects can be minimized by restraining the motion of the atoms in the outer shell of the system, viz., the *extended wall region* [53]. The atoms in this region can be kept fixed or harmonically restrained to stationary positions. Atoms in the inner region are simulated without any such restraints. However, it has not yet been analyzed whether restraining the motion of part of the system affects the obtained estimate of ΔF_{BA}. Especially when many atoms are created or annihilated, one would like to let the system adapt its volume, that is, simulate at constant (T,P) instead of having the volume more or less fixed by the extended wall region of positionally restrained atoms. In any case, the treatment of the boundary in both processes 3 and 4 in scheme (7.1) should be as similar as possible, so that artifacts of the approximations may cancel when calculating $\Delta F_4 - \Delta F_3$ [30, 54]. If one process (e.g., process 3) is simulated using periodic boundary conditions and the other (e.g., process 4) using an extended wall region of restrained atoms, the difference in boundary conditions is likely to affect the resulting estimate $\Delta F_4 - \Delta F_3$, keeping in mind the long-range character of the Coulomb forces.

8.14. Sensitivity of free energy to force field parameters

Not unexpectedly, the free energy of a molecular system is rather sensitive to the interaction potential or force field (3.1) that is used in the calculation. For example, in the relatively simple case of the free energy of solvation of methanol in water, application of force field parameters from established force fields yields a spread in free energies of solvation of about 2 kT around G_{exp}. This means that the calculation of free energy of solvation of small molecules is a great opportunity to test current force fields by comparison with experimental data and to obtain indications of possible improvement of force field parameters. However, it also means that in practical applications it is advisable to check results with respect to variations in the force field parameters.

8.15. Effect of inclusion of polarizability on free energy estimates

The term in the interaction function (3.1) representing the nonbonded interactions consists only of a summation over all pair interactions in the molecular system. Nonbonded many-body interactions are neglected. Yet, inclusion of polarizability of atoms or bonds will be inevitable, if one would like to obtain truly accurate free energy estimates for, for example, the binding of a charged inhibitor to an enzyme, or for catalysis.

9. Perspectives for free energy calculations

In view of the list of approximations given in Section 8, the relative free energy difference calculated using the thermodynamic cycle integration technique and MD computer simulation should be considered as an estimate number in some cases and as a rather accurate prediction in other cases. Its reliability will depend on whether physically sound approximations and assumptions are made. Agreement with experimentally obtained free energy differences is a necessary, but *not* a sufficient condition for having obtained a reliable free energy estimate. In general, one is comparing a few numbers (ΔFs) calculated for a multi-dimensional system using many assumptions, approximations and parameters. One might well obtain good agreement between calculated and measured numbers for the wrong reasons, viz., accidental agreement, compensation of errors or adjustment of parameters. The many different parameters in this type of calculation and the sensitivity of the free energy to force field parameters in particular allow for an adjustment guided by the desired result.

Part of the assumptions and approximations that were discussed in Section 8 have not yet been tested in practice. This should be done in order to enhance the confidence in the predictive power of this type of calculation.

Finally, we note that the extension of the continuous-coupling thermodynamic cycle integration technique to free energy differences between two non-equilibrium states A and B as proposed here, will be useful when studying activated processes.

Acknowledgements

The author is much indebted to Herman Berendsen for a stimulating collaboration over many years. This work was supported in part by the Foundation for Chemical Research (SON) and in part by a NATO Science Fellowship, both under the auspices of the Netherlands Organisation for the Advancement of Pure Research (ZWO). The hospitality of the Department of Pharmaceutical Chemistry of the School of Pharmacy of the University of California at San Francisco is gratefully acknowledged.

The GROMOS (Groningen Molecular Simulation) program library, which can be used to apply the methods discussed, is available at nominal cost from the author.

References

1. Hermans, J. (Ed.) Molecular Dynamics and Protein Structure, Polycrystal Book Service, P.O. Box 27, Western Springs, IL 60558, 1985.
2. Ciccotti, G. and Hoover, W.G. (Eds.) Molecular-dynamics Simulation of Statistical Mechanical Systems (Proceedings of the International School of Physics 'Enrico Fermi', Course 97), North-Holland, Amsterdam, 1986.
3. Beveridge, D.L. and Jorgensen, W.L. (Eds.) Computer Simulation of Chemical and Biomolecular Systems, Ann. N.Y. Acad. Sci., 482 (1986).
4. McCammon, J.A. and Harvey, S.C., Dynamics of Proteins and Nucleic Acids, Cambridge University Press, Cambridge, 1987.
5. Berendsen, H.J.C., van Gunsteren, W.F., Egberts, E. and de Vlieg, J., In Jensen, K.F. and Truhlar, D.G. (Eds.) Supercomputer Research in Chemistry and Chemical Engineering (ACS Symposium Series, No. 353), American Chemical Society, Washington, DC 1987, pp. 106-122.
6. Van Gunsteren, W.F., Mol. Simulation, in press.
7. Berendsen, H.J.C., Postma, J.P.M. and van Gunsteren, W.F., In Hermans, J. (Ed.) Molecular Dynamics and Protein Structure, Polycrystal Book Service, P.O. Box 27, Western Springs, IL 60558, 1985, pp. 43-46.
8. Frenkel, D., In Ciccotti, G. and Hoover, W.G. (Eds.) Molecular-dynamics Simulation of Statistical Mechanical Systems (Proceedings of the International School of Physics 'Enrico Fermi', Course 97), North-Holland, Amsterdam, 1986, pp. 151-188.
9. Mezei, M. and Beveridge, D.L., In Beveridge, D.L. and Jorgensen, W.L. (Eds.) Computer Simulation of Chemicals and Biomolecular Systems, Ann. N.Y. Acad. Sci., 482 (1986) 1.
10. Friedman, H.L., A Course in Statistical Mechanics, Prentice-Hall, Englewood Cliffs, NJ 07632, 1985.
11. Karplus, M. and Kushick, J.N., Macromolecules, 14 (1981) 325.
12. Edholm, O. and Berendsen, H.J.C., Mol. Phys., 51 (1984) 1011.
13. DiNola, A., Berendsen, H.J.C. and Edholm, O., Macromolecules, 17 (1984) 2044.
14. Widom, B., J. Chem. Phys., 39 (1963) 2808.
15. Powles, J.G., Evans, W.A.B. and Quirke, N., Mol. Phys., 46 (1982) 1347.
16. Guillot, B. and Guissani Y., Mol. Phys., 54 (1985) 455.
17. Shing, K.S. and Gubbins, K.E., Mol. Phys., 46 (1982) 1109.
18. Van Gunsteren, W.F. and Berendsen, H.J.C., Groningen Molecular Simulation (GROMOS) Library Manual, Biomos, Nijenborgh 16, 9747 AG Groningen, The Netherlands, 1987.
19. Kirkwood, J.G., J. Chem. Phys., 3 (1935) 300.
20. Zwanzig, R.W., J. Chem. Phys., 22 (1954) 1420.
21. Torrie, G.M. and Valleau, J.P., Chem. Phys. Lett., 28 (1974) 578.
22. Okazaki, S., Nakanishi, K., Touhara, H. and Adachi, Y., J. Chem. Phys., 71 (1979) 2421.
23. Postma, J.P.M., Berendsen, H.J.C. and Haak, J.R., Faraday Symp. Chem. Soc., 17 (1982) 55.
24. Tembe, T.L. and McCammon, J.A., Computers & Chemistry, 8 (1984) 281.

25. Jorgensen, W.L. and Ravimohan, C., J. Chem. Phys., 83 (1985) 3050.
26. Mruzik, M.R., Abraham, F.F., Schreiber, D.E. and Pound, G.M., J. Chem. Phys., 64 (1976) 481.
27. Berens, P.H., Mackay, D.H.J., White, G.M. and Wilson, K.R., J. Chem. Phys., 79 (1983) 2375.
28. Mezei, M., Swaminathan, S. and Beveridge, D.L., J. Am. Chem. Soc., 100 (1978) 3255.
29. Postma, J.P.M., MD of H_2O: A Molecular Dynamics Study of Water, Ph.D. Thesis, University of Groningen, Groningen, The Netherlands, 1985.
30. Van Gunsteren, W.F. and Berendsen, H.J.C., The power of dynamic modelling of molecular systems, to be submitted to J. Comput.-Aided Mol. Design.
31. Straatsma, T.P., Berendsen, H.J.C. and Postma, J.P.M., J. Chem. Phys., 85 (1986) 6720.
32. Straatsma, T.P., Free Energy Evaluation by Molecular Dynamics Simulations, Ph.D. Thesis, University of Groningen, Groningen, The Netherlands, 1987.
33. Straatsma, T.P. and Berendsen, H.J.C., J. Chem. Phys., 89 (1988) 5876.
34. Van Gunsteren, W.F. and Berendsen, H.J.C., J. Comput.-Aided Mol. Design, 1 (1987) 171.
35. Fleischman, S.H. and Brooks, C.L., J. Chem. Phys., 87 (1987) 3029.
36. Mezei, M., J. Chem. Phys., 86 (1987) 7084.
37. Warshel, A., Sussman, F. and King, G., Biochemistry, 25 (1986) 8368.
38. McQuarrie, D.A., Statistical Mechanics, Harper & Row, New York, NY, 1976.
39. Goldstein, H., Classical Mechanics, Addison-Wesley, Reading, MA, 1950.
40. Berendsen, H.J.C. and Van Gunsteren, W.F., In Perram, J.W. (Ed.) The Physics of Superionic Conductors and Electrode Materials (NATO ASI Series B92), Plenum Press, New York, NY, 1983, pp. 221-240.
41. Berendsen, H.J.C. and van Gunsteren, W.F., In Barnes, A.J., Orville-Thomas, W.J. and Yarwood, J. (Eds.) Molecular Liquids: Dynamics and Interactions (NATO ASI Series C135), Reidel, Dordrecht, 1984, pp. 475-500.
42. Pear, M.R. and Weiner, J.H., J. Chem. Phys., 71 (1979) 212.
43. Chandler, D. and Berne, B.J., J. Chem. Phys., 71 (1979) 5386.
44. Van Gunsteren, W.F., Mol. Phys., 40 (1980) 1015.
45. Ryckaert, J.-P., Ciccotti, G. and Berendsen, J.H.C., J. Comput. Phys., 23 (1977) 327.
46. Singh, U.C., Brown, F.K., Bash, P.A. and Kollman, P.A., J. Am. Chem. Soc., 109 (1987) 1607.
47. Cross, A.J., Chem. Phys. Lett., 128 (1986) 198.
48. Rao, S.N., Singh, U.C., Bash, P.A. and Kollman, P.A., Nature, 328 (1987) 551.
49. Bash, P.A., Singh, U.C., Langridge, R. and Kollman, P.A., Science, 236 (1987) 564-568.
50. Born, M., Z. Phys., 1 (1920) 45.
51. Berendsen, H.J.C., In Hermans, J. (Ed.) Molecular Dynamics and Protein Structure, Polycrystal Book Service, P.O. Box 27, Western Springs, IL 60558, 1985, pp. 18-22.
52. Berendsen, H.J.C., Postma, J.P.M., van Gunsteren, W.F., DiNola, A. and Haak, J.R., J. Chem. Phys., 81 (1984) 3684.
53. Van Gunsteren, W.F. and Berendsen, H.J.C., In Hermans, J. (Ed.) Molecular Dynamics and Protein Structure, Polycrystal Book Service, P.O. Box 27, Western Springs, IL 60558, 1985, pp. 5-14.
54. Berendsen, H.J.C., Comput. Phys. Commun., 44 (1987) 233.

Free energies in solution: The aqua vitae of computer simulations

William L. Jorgensen
Department of Chemistry, Purdue University, West Lafayette, IN 47907, U.S.A.

Introduction

Transformations of molecular systems in solution are the central focus of chemical and biochemical sciences. The connection to processes of life itself is clear and emphasizes the importance of obtaining an atomic level understanding of structure and reactivity in solution. It would obviously be exciting to be able to computationally model complex molecular systems accurately enough to provide valid insights that yield enhanced predictive capabilities for structure and reactivity. The key item that is needed is the ability to calculate free energy changes for equilibria and reactions. In fact, progress is being made along these lines via molecular dynamics and statistical mechanics simulations for organic and biomolecular systems in solution. Summaries of contributions by different groups are available in two recent books [1, 2]. In the present paper, the focus will be on recent results from our group on organic transformations, specifically, solvent effects on two important classes of processes, conformational equilibria and ion-pair complexation. Our work on free energy surfaces for organic reactions in solution that involve changes in covalent bonding has been reviewed recently elsewhere [3].

Methodology

The principal objective of obtaining free energy surfaces for the transformations in solution is clearly challenging if the systems are going to be modeled at the atomic level [3]. The full, multi-dimensional problem for even moderate-sized systems is beyond most available computer resources. Therefore, the work so far has concentrated on systems where a one-dimensional reaction coordinate can be defined. This may be adequate for a simple dissociation process or an internal rotation about one dihedral angle, while for more general transformations a mapping of the geometric changes to one dimension may be possible. For example, in our studies of S_N2 and addition reactions [4, 5], the minimum energy reaction path (MERP), $E^\circ(r_c)$, for the gas phase was determined as a function of a single geometrical variable, such as a bond length, which is taken as the reaction coordinate r_c. Of course, there are other geometrical changes that occur as a function of r_c and that are expressed as smooth functions, $P_i(r_c)$, for each variable i. The effect of solvation on the MERP with its multi-dimensional geometrical changes can then be computed as a function of r_c, though the possibility of a change in mechanism (MERP) in solution must be considered [6]. The interconversion of two conformational isomers might also involve changes in multiple dihedral angles, e.g., Φ and Ψ for a peptide, that, however, can be mapped to a one-dimensional, possibly arbitrary reaction path for computation of the relative solvation energies for the conformers [7].

Once the reaction or transformation path has been defined, two approaches have been taken to evaluating the free energy profile in solution, $W(r_c)$. The simpler one conceptually is to use statistical perturbation theory to 'step along' the reaction coordinate and compute the free energy change at each step. The procedure follows from Eq. 1 [8]

$$G_j - G_i = -k_B T \ln \langle \exp[-\beta(E_j - E_i)] \rangle_i \quad (1)$$

which represents the free energy difference between points i and j along r_c as a function involving the energy difference for the two points. In the equation, $\beta = (k_B T)^{-1}$ and the average is based on running the simulation in the isothermal-isobaric ensemble for the system at point i. Thus, j is treated as a perturbation from i, and, as usual, the perturbation cannot be too large or convergence of Eq. 1 will be too slow. Typical differences in i and j might be 0.1-0.2 Å for a distance and 15° for a dihedral angle; however, twice these values, i.e., 0.2-0.4 Å and 30°, can be covered in one simulation by considering the perturbations from point i to i–1 and to i+1.

The alternative procedure involves a series of simulations with 'importance sampling' [3-5, 9] that each cover a range of r_c and yield $W(r_c)$ continuously. Biasing functions are used to constrain the system to a range around a point r_c^i. The reaction coordinate is then allowed to vary during the simulation just like any other variable. The occurrence of different values of r_c is recorded as a distribution function, $g_i(r_c)$, which is simply related to the relative free energy or 'potential of mean force' (pmf) by Eq. 2

$$W(r_c) = -k_B T \ln g_i(r_c) \quad (2)$$

once correction is made for the influence of the biasing. Repetition for overlapping ranges of r_c or 'windows' then allows splicing of the individual $g_i(r_c)$'s to yield $W(r_c)$ for the full range of r_c. The difficulties with this approach include the choice of biasing functions and concern that the sampling in each window is complete.

The remaining critical component in the simulations is the potential functions that describe the intra- and intermolecular energetics for the system. The intermolecular part is usually represented in a Coulomb plus Lennard-Jones form with the interactions occurring between sites located on the nuclei. Simple potential functions are now available that give excellent thermodynamic and structural results for many pure liquids including water [10], hydrocarbons [11], alcohols [12], ethers [13] and sulfur compounds [14]. Force fields for proteins and peptide-water interactions are also well-evolved [15-17]. The biggest problem is developing potential functions for specific reacting systems in which the potential function parameters, e.g., atomic charges, clearly vary along the reaction coordinate. A viable approach is to fit to results of ab initio molecular orbital calculations [4,5]. Other difficulties are associated with the treatment of polarization particularly for highly charged atomic sites and of long-range electrostatic interactions. The former issue is a three-body effect that has so far received little attention in large-scale molecular simulations, while the latter, related problem is particularly serious for ionic solids, molten salts, and polyelectrolytes such as nucleic acids [18].

Given the reaction path and the potential functions, $W(r_c)$ can then be determined via Eq. 1 or 2 using either Monte Carlo or molecular dynamics simulations. Our work on the representative organic systems summarized below has featured Monte Carlo statistical mechanics for the reacting system plus 200-300 solvent molecules in a cubic or rectangular box with periodic boundary conditions. The isothermal-isobaric (NPT) ensemble has been employed at 25°C and 1 atm. The sampling for each step or window in the free energy calculations has typically consisted of an equilibration phase for 10^6 configurations followed by averaging for 2×10^6 configurations and would require ca. 10 days on a computer like a VAX 11/780. Additional details may be found elsewhere [3-6].

Applications

Conformational free energy surfaces

Fluid simulations have been used to investigate the effect of condensed-phase environments on conformational energy surfaces and equilibria for both pure liquids and dilute solutions [19]. In the case of rotation about one bond, the dihedral angle is the reaction coordinate for this simple isomerization. The studies so far have mostly assumed the rotation to be rigid, though concurrent variations in bond lengths and bond angles can easily be accommodated. In the latter case, a simulation for the rotation of an isolated molecule in the gas phase would yield the reference dihedral angle distribution, $S°(\Phi)$, which could be compared to the result in solution, $S(\Phi)$. The two distributions are related to the potentials of mean force for the gas phase, $W°(\Phi)$, and solution, $W(\Phi)$, by Eqs. 3 and 4

$$W°(\Phi) = -k_BT \ln S°(\Phi) + \text{const.} \tag{3}$$

$$W(\Phi) = -k_BT \ln S(\Phi) + \text{const.} \tag{4}$$

which are analogous to Eq. 2. $W°(\Phi)$ and $W(\Phi)$ both contain terms for sampling over the intramolecular vibrations, while $W(\Phi)$ also embodies the effects of any frequency shifts for the vibrations in solution as well as the change in free energy of solvation as a function of Φ, $\Delta G^{sol}(\Phi)$. In the simplified case where the intramolecular vibrations are ignored, $W°(\Phi)$ is just the gas-phase torsional potential, $V(\Phi)$, and $W(\Phi) = V(\Phi) + \Delta G^{sol}(\Phi)$. The solvent effect on $S(\Phi)$ is then clearly seen through Eq. 5 where c is a normalization constant.

$$S(\Phi) = c\, S°(\Phi) \exp(-\beta\, \Delta G^{sol}(\Phi)) \tag{5}$$

Calculation of $S(\Phi)$ for many pure organic liquids such as alkanes, alcohols, ethers, thiols, sulfides, disulfides, and haloalkanes is easily achieved by directly sampling over the dihedral angle(s) or by umbrella sampling using reduced torsional barriers [11-14, 19]. However, dilute solutions, which are modeled as a single solute plus several hundred solvent molecules, present potential convergence problems since statistics are only obtained on the one solute instead of the hundreds of identical molecules in the simulation of a pure liquid, i.e., in a Monte Carlo calculation most of the time would be spent moving the solvent molecules.

Several approaches can be used to address the problem. For one dihedral angle, umbrella sampling over a torsional surface with greatly reduced barriers is viable coupled with long simulations [19, 20]. In Monte Carlo calculations, convergence of $S(\Phi)$ can also be enhanced by preferential sampling whereby moves of the solute and the nearby solvent molecules are attempted more frequently than for the more distant solvent molecules [19, 21]. For example, this approach was used in several simulations of butane in water that were aimed at studying the hydrophobic effect on the trans-gauche equilibrium [19, 22, 23]. Though the potential functions have varied, the results have consistently revealed a 10-20% increase in the gauche population at 25°C upon transfer from the gas phase to water. This is consistent with other theoretical results [24, 25] and fundamental notions about hydrophobic effects on biomolecular structure [26]. Furthermore, the effect has been shown not to just be due to the fluid environment, since no conformational shift is found for pure liquid butane [11, 19].

Statistical perturbation theory has also recently been applied to this problem

[27]. The major advantage to this procedure is that the reaction coordinate, Φ, is progressively stepped along so there is no concern about the thoroughness of sampling over Φ. In addition, the convergence of the calculations was readily checked by running the perturbations from $\Phi = 0°$ to 180° and then back from $\Phi = 180°$ to 0° in 15° increments. The hysteresis amounted to only 1-2% in the conformer populations and a 12% increase in the gauche population was obtained at 25°C using the TIP4P model for water [10] and the OPLS parameters for butane [11]. The results for $S°(\Phi)$ and $S(\Phi)$ are compared in Fig. 1; the trans population decreases from 68% in the gas phase using the MM2 $V(\Phi)$ [22] to 56% in water.

In another recent application of the perturbation method, the interconversion of the trans and cis conformers of *N*-methylacetamide was modeled in TIP4P water at 25°C [28]. The two conformers were found to have the same free energies of hydration. Thus, the free energy difference of 2.5 kcal/mol favoring the trans rotamer in the gas phase also pertains to aqueous solution. The substantial preference for the trans form in water is consistent with the extreme rarity of cis peptide bonds in proteins not involving Pro residues.

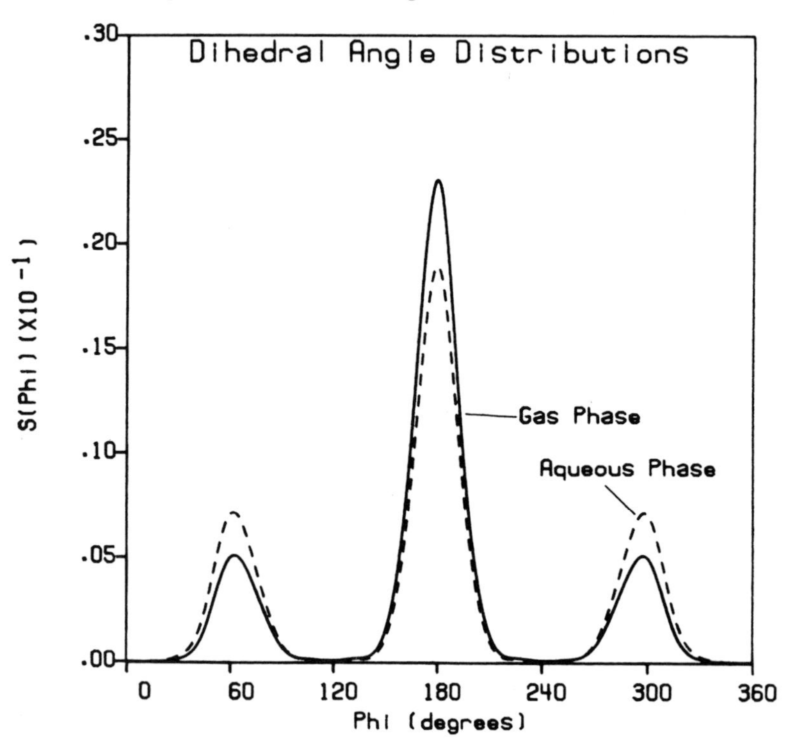

Fig. 1. Population distributions for the central CC bond of butane in the gas phase and in aqueous solution at 25°C. Units for the ordinate are mole fraction per degree.

For more substantial structural differences between conformers involving multiple dihedral angles, importance sampling or the perturbation procedure are probably the best approaches using a one-dimensional reaction path with concerted variations in the angles [7]. Obtaining even a complete two-dimensional map for the solvent effect, e.g., $\Delta G^{sol}(\Phi,\Psi)$, is computationally too demanding. Adequate simulations for the one-dimensional case require $(1\text{-}2)\times 10^7$ Monte Carlo configurations, though the earlier studies with umbrella sampling used less than half this number. The full N-dimensional problem would require ca. 10^{N-1} times the effort for similar grid spacings.

Complexation

Another fundamental class of molecular transformations features the complexation (association or dissociation) of two species. The potential of mean force needs to be determined as a function of the inter-species separation in this case. For two spherical atoms or ions a one-dimensional reaction coordinate is appropriate and $W(r_c)$ will reveal important information, e.g., on the existence of contact and solvent-separated ion pairs and on the intervening barriers. The treatment of non-spherical substrates can either involve some geometrical restrictions on the approach path or include appropriate orientational averaging with r_c as a characteristic distance such as the centers-of-mass separation. Since it will generally be desirable to cover substantial ranges of r_c, importance sampling or statistical perturbation theory can be used to obtain $W(r_c)$.

Few studies have been carried out along these lines in view of the computational demands of the multiple simulations. The initial investigations were for two Lennard-Jones particles or two benzene molecules in water and featured importance sampling [9b, 29]. The results showed the contact free energy minima anticipated from the hydrophobic effect as well as intriguing solvent-separated minima. Importance sampling was also used more recently to compute a pmf for Na^+Cl^- in water, again revealing contact and solvent-separated minima [30]. Then, in the last two years, our group has applied the perturbation method to obtain potentials of mean force for three prototypical organic ion pairs, $(CH_3)_3C^+Cl^-$, $(CH_3)_4N^+Cl^-$, and $CH_3NH_3^+CH_3COO^-$, in water [31, 32].

The *t*-butyl chloride system was examined to help elucidate the energetics in the ion-pair region for a solvolysis reaction. The results have been described in detail elsewhere [31] though the computed pmf is reproduced in Fig. 2. The system consisted of 250 TIP4P water molecules plus the ion pair in a rectangular periodic cell. The chloride ion was constrained to lie on the three-fold axis for the planar *t*-butyl cation, which is the favored orientation in the absence of solvent, and the reaction coordinate was taken as the central carbon-chlorine distance. Perturbations of ± 0.125 or ± 0.25 Å along r_c were found to be acceptable with a total of 15 Monte Carlo simulations used to cover r_c from 2.5 to 8.0 Å.

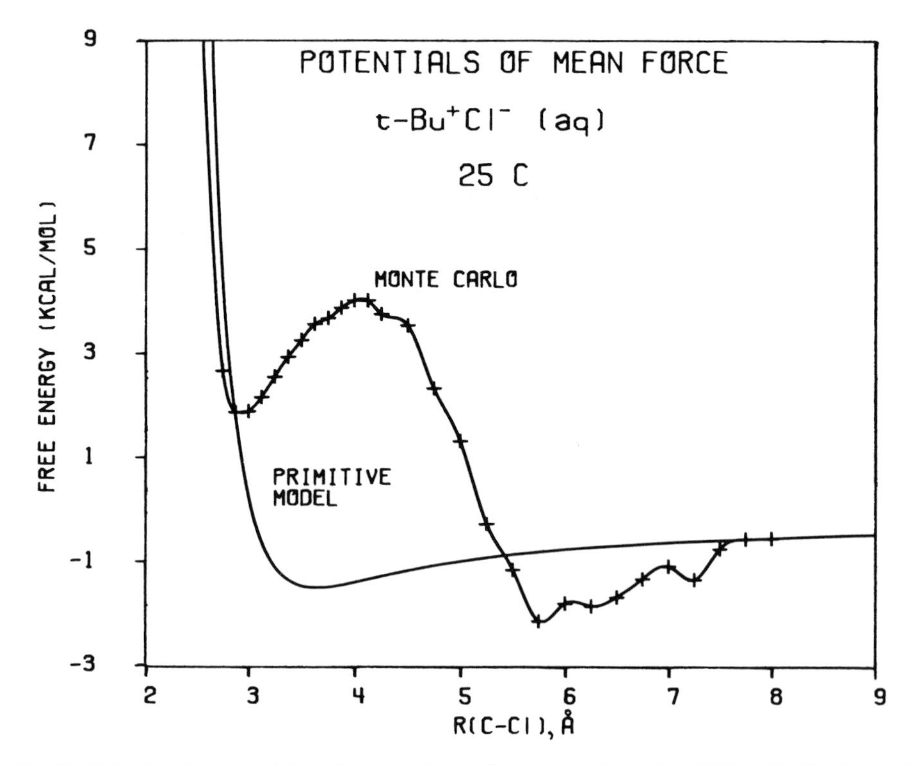

Fig. 2. Calculated potentials of mean force for separating the $(CH_3)_3C^+Cl^-$ ion pair in water at 25°C.

As shown in Fig. 2, minima were found at 2.9 and 5.75 Å corresponding to contact and solvent-separated ion pairs. This is consistent with Winstein's ion-pair scheme, though the simulations revealed no statistically significant minima beyond 5.75 Å which suggests there is just a continuum of structures between the solvent-separated ion pair and 'free ion' stages [31]. The delicacy of the results should be noted since a remarkable cancellation of the ion-ion and ion-solvent interactions occurs to leave the residual pmf. Though the pmf varies by only 5 kcal/mol between 3 and 8 Å, the two components vary by ca. 60 kcal/mol in opposite directions. For comparison, the 'primitive model' prediction is also shown in Fig. 2. It is obtained from the gas-phase ion-ion interaction by dividing the Coulombic term by the experimental dielectric constant, and is also used to anchor the Monte Carlo results at 8 Å. The importance of the explicit structure of the water molecules on the pmf is apparent from comparing the two curves.

The corresponding results for the tetramethylammonium chloride ion pair are shown in Fig. 3 as obtained from analogous Monte Carlo simulations.

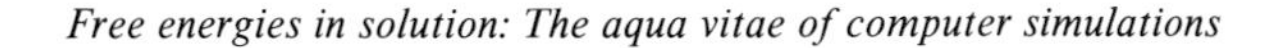

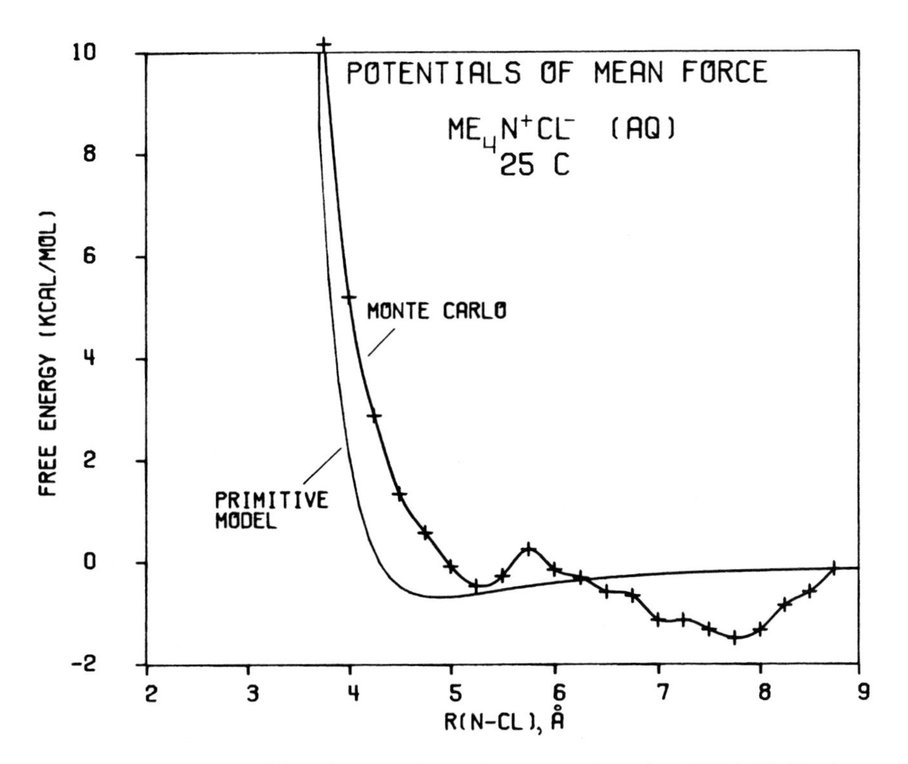

Fig. 3. Calculated potentials of mean force for separating the $(CH_3)_4N^+Cl^-$ ion pair in water at 25°C.

The barrier between the contact and solvent-separated minima is now very small and the deviations from the primitive model are modest. The $(CH_3)_4N^+$-water attractions are particularly weak, <10 kcal/mol [32, 33] and there is little orientational dependence for the interactions in comparison to $(CH_3)_3C^+$. The barrier between the minima may be greater for the carbenium ion due to the orientational preference of having two water molecules or a water molecule and the counterion on either side of the central carbon, which has the largest fractional positive charge (0.4 e) and smallest Lennard-Jones σ [31]. Another interesting feature in Fig. 3 is the location of the contact minimum at an N-Cl separation of 5-5.25 Å. It has been pushed out from the primitive model result by ca. 0.5 Å due most likely to the screening of the chloride ion by the large ammonium ion at short separations. An illustration of the contact ion pair is provided in Fig. 4. Though a water molecule cannot fit directly between the ions at this distance, keeping the ions farther apart helps retain the number of hydrogen bonds to the chloride ion near the 7-8 that are found for the isolated anion in water.

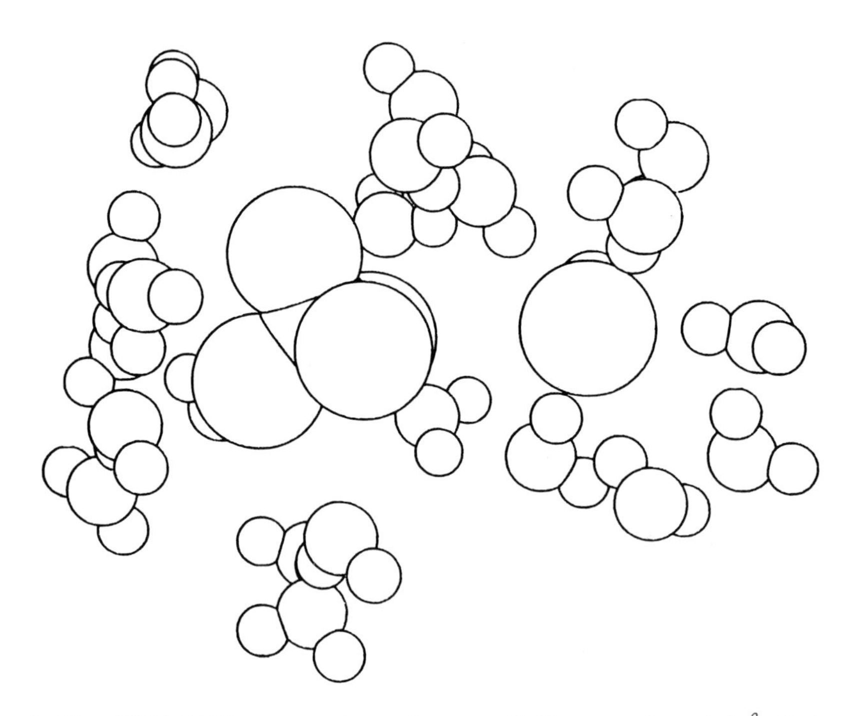

Fig. 4. The $(CH_3)_4N^+Cl^-$ contact ion pair at an N-Cl separation of 5 Å. Only water molecules with oxygens within 4.5 Å of any atom in the ion pair are shown. Taken from an arbitrary configuration in the Monte Carlo simulation.

The methylammonium acetate system was studied as a simple model for salt bridges in proteins. The Monte Carlo calculations were carried out with the ion pair oriented as $CH_3NH_3^+\cdots{}^-O_2CCH_3$ and with the C–N···C–C unit colinear, though rotation of the two ions about this axis was allowed. This approach is the lowest energy one in the absence of solvent and is consistent with the structures observed for salt bridges. The optimal interactions with a water molecule are stronger for both $CH_3NH_3^+$ (18 kcal/mol) and $CH_3CO_2^-$ (16-19 kcal/mol) than for the other ions, $(CH_3)_3N^+$ (9 kcal/mol), $(CH_3)_3C^+$ (16 kcal/mol), and Cl^- (13 kcal/mol) [31, 33]. The increased ion-water attractions are apparently enough to fully offset the increased ion-ion attraction, so the computed pmf continually rises from an N···C separation of 8 Å down to 3 Å, as shown in Fig. 5. Such behavior is not unreasonable for strong electrolytes, though the lack of polarization in the potential functions is always a concern. It is also not inconsistent with the thermodynamics of forming salt bridges which are estimated to only stabilize a protein by 1-2 kcal/mol [34]. The more favorable interaction in proteins follows since the protein backbone constrains the side chains and terminal ionic groups, and the salt bridges are normally formed

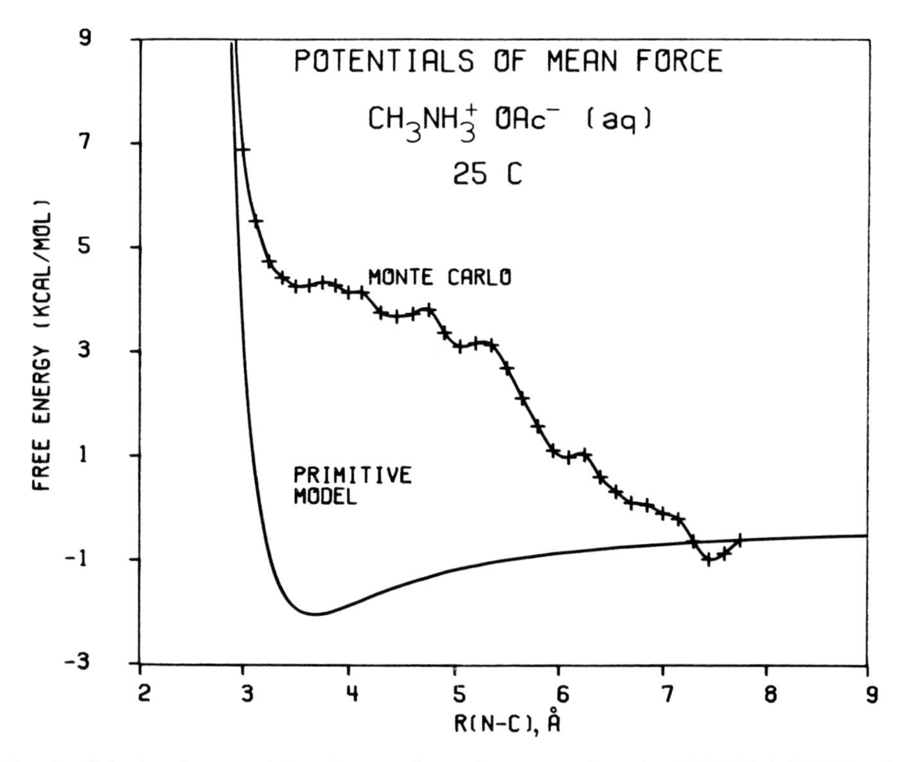

Fig. 5. Calculated potentials of mean force for separating the $CH_3NH_3^+CH_3CO_2^-$ ion pair in water at 25°C.

on the protein surface where the effective dielectric constant is substantially less than for bulk water.

Other work on complexation should be noted, especially the recent computations of relative binding affinities for substrates to enzymes in water [35, 36]. Statistical perturbation theory was used to compute the free energy for mutating

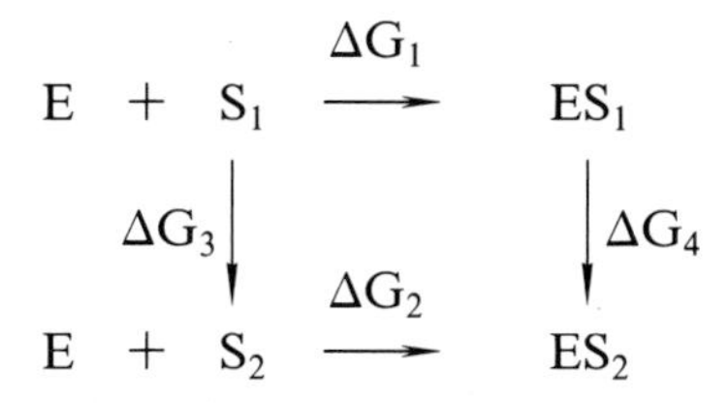

S_1 to S_2 and ES_1 to ES_2 which yields the relative binding affinity from the thermodynamic cycle since $\Delta G_1 - \Delta G_2 = \Delta G_3 - \Delta G_4$.

We note here that an efficient method may be devised for computing absolute binding energies for such complexes or for anchoring the ion-pair pmfs discussed above. Consider the thermodynamic cycle below for the general complexation of A and B:

$$\begin{array}{ccccc} A & + & B & \xrightarrow{\Delta G_{gas}} & AB \\ \Delta G_A \downarrow & & \downarrow \Delta G_B & & \downarrow \Delta G_{AB} \\ A & + & B & \xrightarrow[\Delta G_{sol}]{} & AB \end{array}$$

ΔG_A, ΔG_B and ΔG_{AB} are the free energies of solvation of the respective species from the gas phase and ΔG_{sol} is the desired free energy of complexation in solution. ΔG_{gas}must be known from a quantum or molecular mechanics calculation. Then the free energies of solvation could be obtained by mutating A, B, and AB to nothing in solution, i.e., making them disappear [37]. Thus, Eq. 6 would apply.

$$\Delta G_{sol} = \Delta G_{gas} + \Delta G(0 \rightarrow AB) - \Delta G(0 \rightarrow A) - \Delta G(0 \rightarrow B) \quad (6)$$

However, $0 \rightarrow AB$ can be achieved in two steps as $0 \rightarrow A$ and $A \rightarrow AB$, so $\Delta G(0 \rightarrow AB) = \Delta G(0 \rightarrow A) + \Delta G(A \rightarrow AB)$. This substitution yields Eq. 7

$$\Delta G_{sol} = \Delta G_{gas} + \Delta G(A \rightarrow AB) - \Delta G(0 \rightarrow B) \quad (7)$$

which shows that only two simulations would be needed in both of which B vanishes. Clearly, choosing B to be the smaller component would be computationally desirable. The overall efficiency of the procedure in comparison to Eq. 6 is also evident.

The procedure has been used recently to calculate the absolute free energy of binding for two methane molecules (represented as single Lennard-Jones particles) in TIP4P water at 25°C. The C-C separation was set at 4 Å which corresponds to the contact minimum in the potential of mean force. For this separation, $\Delta G_{gas} = -0.27$ kcal/mol and Monte Carlo simulations yielded $\Delta G_{hyd}(CH_4 \rightarrow (CH_4)_2) = 1.9$ kcal/mol and $\Delta G_{hyd}(0 \rightarrow CH_4) = 2.5$ kcal/mol. Therefore, the computed binding free energy in water is –0.9 kcal/mol with an uncertainty of ca. ± 0.4 kcal/mol. This result is similar to findings from integral equation theories [9b, 38] and the computed ΔG_{hyd} for methane is close to the experimental value of 2.0 kcal/mol [39]. Thus, hydrophobic effects are predicted to favor hydration of the contact methane dimer over the separated species by 0.6 kcal/mol. This is augmented by the intrinsic interaction of 0.3 kcal/mol to yield a net attraction of 0.9 kcal/mol for two methanes in water.

Conclusion

Substantial progress is being made in theoretically modelling organic and biomolecular transformations in solution. The methods reviewed here in conjunction with Monte Carlo and molecular dynamics simulations are versatile and will be applied to an increasing range of problems. The high precision that is now available in computing free energy changes represents a major advance that allows meaningful quantitative predictions as well as direct comparisons between experiment and theory. The thermodynamic results are also complemented by the extreme details on structure and dynamics that are available from the simulations. This is clearly an exciting period during which the atomic-level understanding of the structure and energetics of molecular systems in solution will be greatly enhanced.

Acknowledgements

Gratitude is expressed to the National Science Foundation and National Institutes of Health for support of this research.

References

1. Beveridge, D.L. and Jorgensen, W.L. (Eds.) Computer Simulation of Chemical and Biomolecular Systems, Ann. N.Y. Acad. Sci., 482 (1986) 1.
2. McCammon, J.A. and Harvey, S.C., Dynamics of Proteins and Nucleic Acids, Cambridge University Press, Cambridge, 1987.
3. Jorgensen, W.L., Adv. Chem. Phys., Part II, 70 (1988) 469
4. (a) Chandrasekhar, J., Smith S.F. and Jorgensen W.L., J. Am. Chem. Soc. 106 (1984) 3049 and 107 (1985) 154.
   (b) Chandrasekhar, J. and Jorgensen, W.L., J. Am. Chem. Soc., 107 (1985) 2974.
5. Madura, J.D. and Jorgensen, W.L., J. Am. Chem. Soc., 108 (1986) 2517.
6. Jorgensen, W.L. and Buckner, J.K., J. Phys. Chem., 90 (1986) 4651
7. Mezei, M., Mehrotra, P.K. and Beveridge, D.L., J. Am. Chem. Soc., 107 (1985) 2239.
8. Zwanzig, R.W., J. Chem. Phys., 22 (1954) 1420.
9. (a) Patey, G.N. and Valleau, J.P., J. Chem. Phys., 63 (1975) 2334.
   (b) Pangali, C.S., Rao, M. and Berne, B.J., J. Chem. Phys., 71 (1979) 2975.
10. (a) Jorgensen, W.L., Chandrasekhar, J., Madura, J.D., Impey, R.W. and Klein, M.L., J. Chem. Phys., 79 (1983) 926.
    (b) Jorgensen, W.L. and Madura, J.D., Mol. Phys., 56 (1985) 1381.
11. Jorgensen, W.L., Madura, J.D. and Swenson, C.J., J. Am. Chem. Soc., 106 (1984) 6638.
12. Jorgensen, W.L., J. Phys. Chem., 90 (1986) 1276.

13. (a) Jorgensen, W.L. and Ibrahim, M., J. Am. Chem. Soc., 103 (1981) 3976.
    (b) Briggs, J.M. and Jorgensen, W.L., to be published.
14. Jorgensen, W.L., J. Phys. Chem., 90 (1986) 6379.
15. For a review, see: Levitt, M., Annu. Rev. Biophys. Bioeng., 11 (1982) 251.
16. Weiner, S.J., Kollman, P.A., Case, D.A., Singh, U.C., Ghio, C., Alagona, G., Profeta, S. and Weiner, P., J. Am. Chem. Soc., 106 (1984) 765.
17. (a) Jorgensen, W.L. and Swenson, C.J., J. Am. Chem. Soc., 107 (1985) 569, 1489.
    (b) Jorgensen, W.L. and Tirado-Rives, J., J. Am. Chem. Soc., 110 (1988) 1657.
18. Adams, D.J., Chem. Phys. Lett., 62 (1979) 329.
19. For a review, see: Jorgensen, W.L., J. Phys. Chem., 87 (1983) 5304.
20. Rebertus, D.W., Berne, B.J. and Chandler, D., J. Chem. Phys., 70 (1979) 3395.
21. Owicki, J.C., ACS Symp. Ser., 86 (1978) 159.
22. Jorgensen, W.L., J. Chem. Phys., 77 (1982) 5757.
23. Jorgensen, W.L., Gao, J. and Ravimohan, C., J. Phys. Chem., 89 (1985) 3470.
24. Pratt, L.R. and Chandler, D., J. Chem. Phys., 67 (1977) 3683.
25. Rosenberg, R.O., Mikkilineni, R. and Berne, B.J., J. Am. Chem. Soc., 104 (1982) 7647.
26. Ben-Naim, A., Hydrophobic Interactions, Wiley, New York, 1980.
27. Jorgensen, W.L. and Buckner, J.K., J. Phys. Chem., 91 (1987) 6083.
28. Jorgensen, W.L. and Gao, J., J. Am. Chem. Soc., 110 (1988) 4212.
29. (a) Ravishanker, G., Mezei, M. and Beveridge, D.L., Faraday Symp. Chem. Soc., 17 (1982) 79.
    (b) Ravishanker G. and Beveridge, D.L., J. Am. Chem. Soc., 107 (1985) 2565.
30. Berkowitz, M., Karim, O.A., McCammon, J.A. and Rossky, P.J., Chem.Phys. Lett., 105 (1984) 577.
31. Jorgensen, W.L., Buckner, J.K., Huston, S.E. and Rossky, P.J., J. Am. Chem. Soc., 109 (1987) 1891.
32. Buckner, J.K. and Jorgensen, W.L., submitted for publication.
33. Jorgensen, W.L. and Gao, J., J. Phys. Chem., 90 (1986) 2174.
34. Perutz, M.F., Nature 228 (1970) 726.
35. Wong, C.F. and McCammon, J.A., J. Am. Chem. Soc., 108 (1986) 3830.
36. Bash, P.A., Singh, U.C., Langridge, R. and Kollman, P.A., Science, 236 (1987) 564.
37. Straatsma, T.P., Berendsen, H.J.C. and Postma, J.P.M., J. Chem. Phys., 85 (1986) 6720.
38. Pratt, L.R. and Chandler, D., J. Chem. Phys., 67 (1977) 3683.
39. Yaacobi, M. and Ben-Naim, A., J. Phys. Chem., 78 (1974) 175.

Thermodynamic calculations on biological systems

Charles L. Brooks III
Department of Chemistry, Carnegie Mellon University,
Pittsburgh, PA 15213, U.S.A.

Introduction

Computational methods applied to the study of protein structure and dynamics have provided detailed atomic information about fluctuations of biological molecules and the potential role of these motions in function [1]. In addition to the insights gained into the dynamic behavior of biological macromolecules, simulation techniques are beginning to show some promise in evaluating and rationalizing thermodynamic properties, e.g., relative and absolute free energies of solution and transfer [2]. Initial applications of thermodynamic simulation (TS) methods to the calculation of relative free energies of hydration for small solutes [3,4], relative transfer free energies for amino acids and nucleotides [5], and the binding properties of small molecules to receptor sites [6,7] (in proteins, DNA or synthetic systems) provide encouraging results which suggest this approach will be useful in the rationalization, prediction and design of drugs and enzymes. However, much work remains to be done before the computational capabilities and accuracy of these methods are fully developed and are applicable on a routine basis.

This paper deals with new implementations of the thermodynamic simulation methods (TS) in the program CHARMM [8]. The general formalism for these techniques follows from early work by Zwanzig [9] but takes on several guises in present implementations. The two approaches on which we focus our discussions are termed thermodynamic cycle perturbation theory (TP) [10] and thermodynamic integration (TI) [2].

Several applications of these techniques will be illustrated to demonstrate the potential scope of the methodology. We shall first discuss relative solvation-free energies of alkanes and alcohols in aqueous solution. In this application the focus will be on the use of TS methods to explore the accuracy of current empirical potential energy functions in representing experimental relative free energies of hydration [3] - it is worth noting that most empirical potentials were not parameterized with free energies in mind. Next, initial applications of TS methods to the rationalization and prediction of drug binding studies on congeners of trimethoprim to the enzyme dihydrofolate reductase from chicken are presented [7]. Questions regarding the sampling and convergence of thermodynamic results will be put forward and discussed. Finally, we shall turn to new methods which are under development. The first applications of TS methods to compute thermodynamic derivative properties [11] and free energy surfaces [12] will be outlined.

Formal Developments

The methods of thermodynamic simulation, as we have already seen from Beveridge and DiCapua's article (see pp. 1-26 of this volume), are based on the definition of a hybrid Hamiltonian which represents some admix of the initial state (1) and final state (2) of the system [2]. Let λ represent the 'coordinate' describing the pathway by which the two systems are interconverted, then the hybrid Hamiltonian may be defined by [3,7,8]

$$H(\lambda) = (1-\lambda)^n H_1 + \lambda^n H_2 \tag{1}$$

where n describes the degree of non-linearity for the thermodynamic pathway. An example, which will be discussed below, may serve to illustrate this point. If one is interested in the relative free energy of solvation of methanol versus ethane in water, a hybrid Hamiltonian is one in which the methyl-hydroxy group is slowly replaced (with a dependence on λ as given above) by a methyl moiety. For a fixed value of λ this system is unphysical but describes a specific thermodynamic system with a Hamiltonian given in Eq. 1 above. The phase space for this system may be sampled using molecular dynamics by integrating the Hamiltonian (or equivalently, the Newtonian) equations of motion

$$\dot{p}_\lambda = \frac{\partial H(\lambda)}{\partial q_\lambda} \tag{2a}$$

$$\dot{q}_\lambda = \frac{\partial H(\lambda)}{\partial p_\lambda} \tag{2b}$$

where $p_\lambda = q_\lambda$ represent momenta and coordinates for the atoms.

In our calculations we have used molecular dynamics to sample the phase space for the hybrid system (details regarding the individual calculations may be found in the tables). To make connections with thermodynamic properties, the phase space configurations are used to compute the average quantity

$$\Delta A(\lambda,\lambda+\Delta\lambda) = -k_B T \ln \left< \exp\left[\frac{-(H(\lambda+\Delta\lambda) - H(\lambda))}{k_B T} \right] \right>_\lambda \tag{3}$$

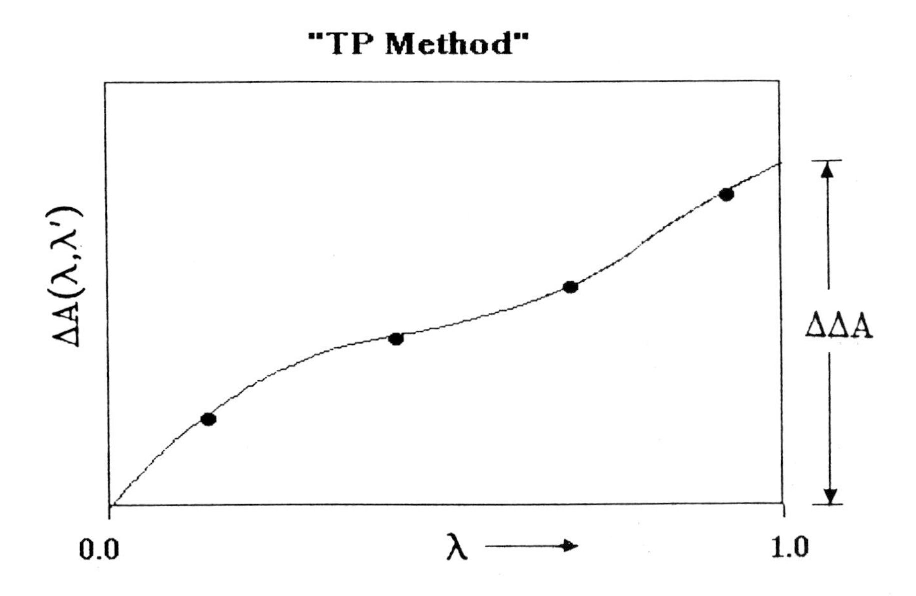

Fig. 1. A schematic representation of results from a free energy simulation using the thermodynamic perturbation (TP) protocol.

which is combined with analogous calculations at different values of λ (different windows). These properties are then summed over the full range of $\lambda=0$ to $\lambda=1$ to compute the free energy change

$$\Delta A = \sum_{\lambda=0}^{1-\Delta\lambda} \Delta A(\lambda,\lambda+\Delta\lambda) \tag{4}$$

A schematic of the free energy changes between 'windows' along a particular thermodynamic pathway is illustrated in Fig. 1.

Alternately, one may use the methods of thermodynamic integration to compute the free energy change. These techniques rely on the connection formula

$$\Delta A = \int_0^1 \left\langle \frac{\partial \Delta A(\lambda)}{\partial \lambda} \right\rangle_\lambda d\lambda \tag{5}$$

where, again, molecular dynamics may be used to compute the average quantities within the broken brackets, $\langle ... \rangle_\lambda$. In most cases this average is calculated for a number of windows spanning $\lambda=0$ to $\lambda=1$ and numerical quadrature

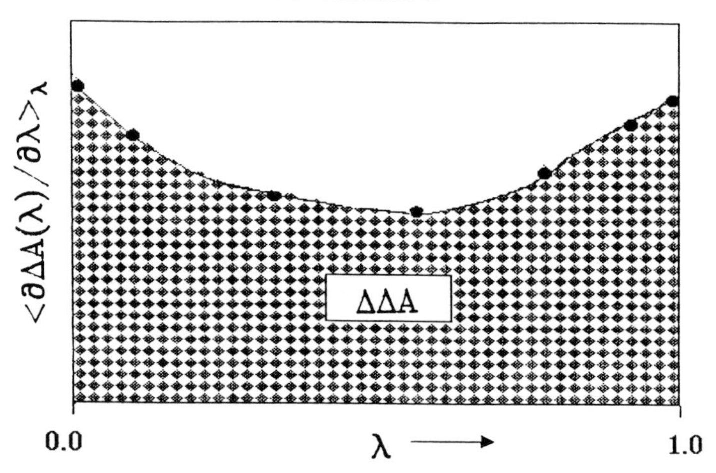

Fig. 2. A schematic representation of results from a free energy simulation using the thermodynamic integration (TI) protocol.

is used to accumulate the free energy change [2,8]. A schematic of the graphical output for the thermodynamic integration analysis is illustrated in Fig. 2.

We have gone beyond the free energy in our work to also make the decomposition into energetic and entropic contributions. One may compute these properties from the same set of thermodynamic simulations used to derive the free energy using the connection formula

$$\Delta E(\lambda,\lambda+\Delta\lambda) \cong \frac{k_B T^2}{2\Delta T} \ln \left[\frac{<\exp\left[\frac{-H(\lambda+\Delta\lambda)}{k_B(T+\Delta T)} + \frac{H(\lambda)}{k_B T}\right]>_{\lambda,T}}{<\exp\left[\frac{-H(\lambda+\Delta\lambda)}{k_B(T-\Delta T)} + \frac{H(\lambda)}{k_B T}\right]>_{\lambda,T}} \right.$$

$$\left. \times \frac{<\exp\left[\frac{-H(\lambda)}{k_B}\left(\frac{1}{T-\Delta T} - \frac{1}{T}\right)\right]>_{\lambda,T}}{<\exp\left[\frac{-H(\lambda)}{k_B}\left(\frac{1}{T+\Delta T} - \frac{1}{T}\right)\right]>_{\lambda,T}} \right] \qquad (6)$$

and

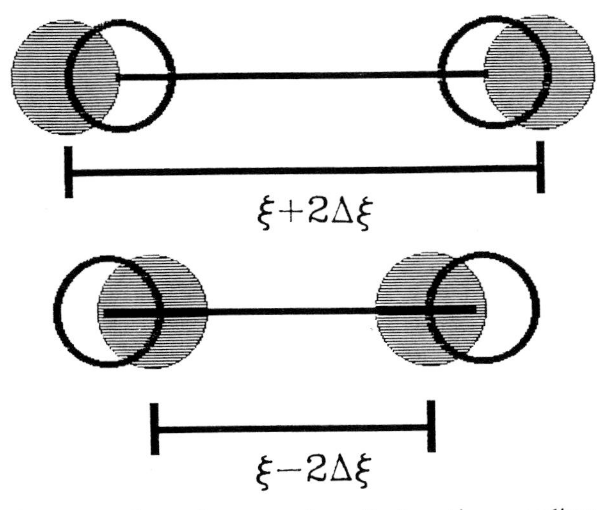

Fig. 3. The displacement of atomic center along the 'reaction coordinate' in calculations of free energy surfaces using Eq. 10.

$$\Delta S(\lambda,\lambda+\Delta\lambda) =$$

$$k_B \ln < \exp\left[\frac{-(H(\lambda + \Delta\lambda) - H(\lambda))}{k_B T}\right] >_{\lambda,T} + \frac{\Delta E(\lambda,\lambda + \Delta\lambda)}{T} \tag{7}$$

where the expressions above are developed by considering the finite difference approximant to the temperature derivative of $\Delta A(\lambda,\lambda+\Delta\lambda)$ [3,7,11]. The total energy and entropy changes are computed by summing incremental changes between each window as done for ΔA in Eq. 4.

Finally, it may be noted that we have also been using thermodynamic simulation methods to calculate *free energy surfaces* using molecular dynamics [12]. Our approach to this calculation represents a fusion of molecular dynamics and Monte Carlo techniques. The basic connection formula follows from Eq. 3 above, and consideration of the free energy difference between a particular configuration specified at ξ and that at $\xi+\Delta\xi$. This is illustrated in Eqs. 8 and 9 and Fig. 3 for the simplest condensed phase reaction coordinate, the separation between two spherical particles. If we let $W_i^{\pm}$ represent the weight (Monte Carlo Boltzmann factor) for displacing the pair of particles by $\pm\Delta\xi$, respectively, during the *i*th molecular dynamics step, as indicated below

$$W_i^{+} = \exp\left\{\frac{-(U(\xi+\Delta\xi) - U(\xi))}{k_B T}\right\} \tag{8}$$

and

$$W_i^- = \exp\left\{\frac{-(U(\xi - \Delta\xi) - U(\xi))}{k_B T}\right\} \tag{9}$$

the free energy difference between the configuration at ξ and those at $\xi \pm \Delta\xi$ is simply related to the average of the weights for each step [12]

$$\Delta\Delta A(\xi,\xi \pm \Delta\xi) = \Delta A(\xi \pm \Delta\xi) - \Delta A(\xi) = \frac{1}{N}\sum_{i=1}^{N} W_i^{\pm}$$

$$= -k_B T \ln\left\langle \exp\left\{\frac{-(U(\xi \pm \Delta\xi) - U(\xi))}{k_B T}\right\}\right\rangle_{\xi} \tag{10}$$

where N is the number of dynamics steps used in computing the average and the subscript ξ on the angle brackets indicates this average is over an ensemble constrained at a fixed ξ.

This formalism is implemented by attempting (but never carrying out) Monte Carlo steps $\pm\Delta\xi$ away from the position ξ and storing the weights for each step in a conventional molecular dynamics simulation. A series of simulations with different ξ spanning the range of reaction coordinate of interest may then be used to compute the reaction free energy surface [12]. Of course, temperature and spatial derivative approximants of $\Delta\Delta A(\xi)$ may be computed in analogy to Eqs. 6 and 7 above. Thus, this method is ripe for exploitation in understanding condensed phase reactivity.

Aqueous Solvation Thermodynamics of Small Molecules [3]

We have explored the calculation of relative thermodynamic properties for the aqueous solvation of small hydrocarbons and alcohols with at least three objectives in mind:

(i) to develop an optimal protocol for carrying out TS calculations,

(ii) to explore the ability of available empirical potential energy functions to accurately reproduce thermodynamic properties from experiment [13],

(iii) to examine the practicability of using these methods as a tool in optimizing empirical energy functions.

Of most importance in these objectives is the second one which focuses on the accuracy of available empirical potentials. It is an essential step in advancing our understanding of the thermodynamic forces underlying protein solvation, folding and protein-drug interactions to know the limits of our mathematical models (i.e., the potentials). The calculations summarized in Table 1 below represent a modest initial step toward this end. We have examined the relative thermodynamics of hydration for ethane, propane and butane (hydrophobic solutes), and methanol and ethanol (polar solutes) using several different model potential functions and two different TS protocols.

Table 1 *Aqueous solvation thermodynamics of alcohols and alkanes*[a]

System and potential function	ΔΔA (kcal/mol)[b]			
	Aqueous	Vacuum	Total	Experiment[c]
Propane → ethane[d]	0.83 ± 0.1	-	0.83 ± 0.1 (2.5 ± 5)	-0.13 (0.65)
Propane → ethane[e]	1.63 ± 0.3	0.75	0.87 ± 0.3 (0.7 ± 6)	-0.13 (0.65)
Propane → ethane[f]	0.56 ± 0.1	-	0.56 ± 0.1 (0.42 ± 4)	-0.13 (0.65)
Butane → propane[f]	-1.08 ± 0.2	-	-1.08 ± 0.2 (7.1 ± 7)	-0.13
Methanol → ethane[f]	7.5 ± 0.2	-	7.5 ± 0.2 (7.1 ± 7)	6.9 (6.1)
Ethanol → propane[f]	5.9 ± 0.2	-	7.1 ± 0.2 (5.6 ± 6)	7.0 (7.2)
Ethanol → propane[g,f]	6.3 ± 0.4	-	7.5 ± 0.4	7.0 (7.2)
Ethanol → propane[h,f]	6.0 ± 0.1	-	7.2 ± 0.1	7.0 (7.2)

[a] All simulations were carried out using 120-124 TIP3P water molecules and one solute at $\rho = 1$ g/cc and $T = 298 \pm 5$ K. The calculations used double-wide sampling with windows at $\lambda = (0.125, 0.5, 0.875)$. Equilibration consisted of 8 ps of equilibration dynamics at each λ and production runs were for 12-20 ps at each λ with a timestep of 1 fs. Estimated errors are $\pm\sigma$.

[b] Numbers in parentheses are energies of hydration, ΔΔE.

[c] Experimental estimates from Ref. 13.

[d] Potential functions for alkanes are from the CHARMM parameter set, see Ref. 14.

[e] Potential functions from all hydrogen parameter set, see Ref. 15.

[f] Potential functions from Jorgensen OPLS parameter set, see Ref. 16.

[g] Slow growth method over period of 10 ps, errors were estimated from forward and reverse calculations.

[h] Slow growth method over period of 20 ps, errors were estimated from forward and reverse calculations.

Examining first the results for apolar → apolar transformations, it is interesting to note that of all three potential functions we examined, two extended hydrocarbon models (CHARMM19 [14] and OPLS [16] and one explicit hydrogen model (a CHARMM parameter set currently under development [15]), give very similar values for the relative free energy of hydration for the propane → ethane transformation. Furthermore, we find that this value of $\Delta\Delta A_{solv} \approx 0.75$ kcal/mol is qualitatively incorrect in the sign, $\Delta\Delta G^{exp}_{solv} = -0.13$ kcal/mol. Although the numbers being compared are small, the statistical significance of the difference between the experimental and computed numbers is at the 90% level [17]. Thus, we must conclude that the three potential models explored in this study are lacking in their accurate representation of apolar thermodynamics.

We have also examined the relative thermodynamics for the transformation of butane → propane, with butane fixed in the trans conformation, using the OPLS parameter set. The sign now changes for $\Delta\Delta A_{solv}$ and is consistent with the experimental number, however, there is still a significant discrepancy between the two. This difference is expected to become even greater when adequate sampling of all conformations is made. We can estimate, using the present value of $\Delta\Delta A_{solv}$ and the earlier calculations by Jorgensen and coworkers on the gauche-trans populations and free energy differences for butane in water [18], the overall relative free energy change and find it to be $\Delta\Delta A^{est}_{solv} \cong -1.3$ kcal/mol. The free energy change is now of the correct sign but too large. This suggests a sensitivity of the free energies for these systems on the values of the van der Waals parameters, in particular, the 'size' of the hydrophobic group appears to be important.

Next we examined polar → apolar transformations for methanol → ethane and ethanol → propane. In both of these cases acceptable agreement with experiment is found. Thus, since these calculations are dominated by the change in electrostatic forces, we conclude that the potential functions being used are sufficiently accurate in representing the thermodynamics of gross electrostatic changes. We also note that one set of these calculations was performed using the 'slow growth' methodology, which is an approximate formulation of the TI method (see Eq. 5) and the agreement with the TP calculations is good.

In summary of these calculations, we have found the following:

(i) the parameterization of current empirical potential functions appears to be inadequate in representing apolar → apolar thermodynamics in aqueous solution,

(ii) polar → apolar transformations are governed by the loss of large electrostatic components and are adequately represented,

(iii) the 'slow growth' approximation to Eq. 5 appears to provide similar results to those from the TP method for this set of calculations.

Finally, we note that TS methods may be an additional tool for use in parameterization of empirical energy functions.

Thermodynamics of Protein-Drug Interactions [7]

The premier goal in the development of TS methods is their use in the rationalization, prediction and design of potentially new compounds (drugs) which possess specific characteristics with respect to their interactions with proteins. The assault on this problem has two major components, one is characterization of the desolvation thermodynamics of the drug, which must accompany the overall binding process; the other is the thermodynamic differences between two drugs bound to a common receptor site. We have addressed the former with simple examples in the previous section and will comment on the

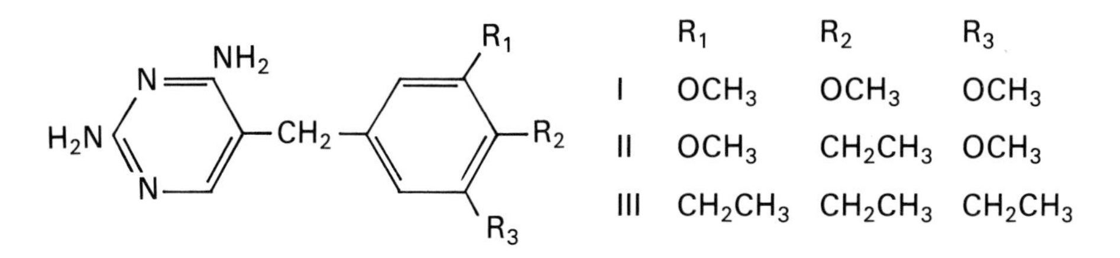

Fig. 4. Structures of trimethoprim (I), and its para-ethyl (II) and tri-ethyl-substituted congeners.

specific compounds of interest below. We then examine the protein-drug interaction thermodynamics and the overall binding process.

Relative solvation thermodynamics

We are beginning to explore the interactions of trimethoprim and a number of congeners with an aqueous environment and with the cofactor-protein complex of NADPH-dihydrofolate reductase (from chicken) [19]. The specific drugs of interest are diagrammed in Fig. 4; we have called them TMP (I), PET (II) and TET (III).

Calculations of the relative thermodynamics of hydration have been carried out for these molecules as well as the protonated relatives with the proton attached to the N1 atom in the pyrimidine ring. The results for these calculations are summarized in Table 2.

With respect to the relative thermodynamics of solvation for the unprotonated drugs we note the following points of interest. First, the transformation which replaces all of the methoxy moieties on the benzyl ring with ethyl groups is dominated by energetics, a result consistent with our earlier calculations of polar to apolar transformations [3]. This is not surprising since we are replacing three potentially hydrogen bonding (polar) groups by apolar groups. When these results are compared with the transformation TMP → PET, however, a number of interesting observations emerge.

(i) The relative free energy changes between the TMP → PET and TMP → TET transformations do not represent additive increments in free energy.

(ii) This is also true for the energies, in fact the thermal decomposition of $\Delta\Delta A_{solv}$ for TMP → PET is dramatically different than for TMP → TET. As we shall see below, this difference contributes to a stronger binding of PET to DHFR (a picture which opposes what might be predicted based solely on desolvation thermodynamics).

(iii) We have examined a slightly modified parameter set for the TMP → TET transformation, with larger van der Waals radii on the ring carbons, and find that the overall thermodynamics of solvation are similar.

Table 2 *Aqueous solvation thermodynamics of drug molecules*[a]

System		ΔΔA (kcal/mol)	ΔΔE (kcal/mol)	ΔΔS (eu)
TMP → TET	(s)	3.8 ± 0.5	4.1 ± 11	0.8 ± 37
	(a)	-4.9 ± 0.3	-2.7 ± 6	7.5 ± 21
	(v)	-8.8 ± 0.4	-6.8 ± 9	6.7 ± 31
TMP → TET[b]	(s)	4.5 ± 0.5	5.3 ± 17	2.4 ± 60
	(a)	8.7 ± 0.4	11.1 ± 11	8.0 ± 38
	(v)	4.1 ± 0.1	5.8 ± 13	5.6 ± 46
TMP → PET	(s)	1.9 ± 0.2	-11.4 ± 5	-44.7 ± 18
	(a)	-1.5 ± 0.2	-16.5 ± 5	-50.5 ± 16
	(v)	-3.4 ± 0.1	-5.2 ± 2	-5.8 ± 16
TMP → $TMPH^+$	(s)	-31.1 ± 0.8	0.1 ± 16	104 ± 54
	(a)	-12.8 ± 0.8	18.2 ± 15	104 ± 50
	(v)	18.2 ± 0.3	18.1 ± 6	-0.3 ± 20
TET → $TETH^+$	(s)	-30.2 ± 0.6	-15.7 ± 13	49 ± 43
	(a)	-13.9 ± 0.6	-0.2 ± 13	46 ± 42
	(v)	16.3 ± 0.1	15.5 ± 2	-2.7 ± 7
PET → $PETH^+$	(s)	-30.6 ± 0.4	-21.7 ± 10	31.5 ± 33
	(a)	-12.8 ± 0.4	-4.6 ± 8	27.5 ± 28
	(v)	17.8 ± 0.2	17.1 ± 5	-2.7 ± 17

[a] All simulations were carried out in 196 TIP3P water molecules [20] at $\rho = 1$ g/cc and T = 298 ± 5 K using a timestep of 1.5 fs. Double-wide sampling at values of $\lambda = (0.125, 0.5, 0.875)$ was used for simulations TMP → TET and TMP → PET, and the calculations consisted of 20 ps equilibration and 30 ps production; double-wide sampling with $\lambda = 0.5$ was used for TMP → $TMPH^+$, TET → $TETH^+$ and PET → $PETH^+$ and the calculations consisted of 15 ps equilibration and 30 ps production simulations using parameters from CHARMM (see Refs. 7 and 14).

Abbreviations: TMP = trimethoprim; TET = tri-ethyl-substituted congener; PET = para-ethyl-substituted congener (see Fig. 4). The symbols (s), (a), and (v) denote, respectively, the total free energy of hydration, which is computed from the difference of calculations carried out in aqueous (a) and vacuum (v) environments.

[b] A parameter set with slightly larger van der Waals radii for ring carbons was used here. Note overall ΔΔA cycle is relatively insensitive to changes in parameters but individual components differ.

In comparing the free energies associated with the addition of a proton at the N1 position of the pyrimidine ring (far removed from the other substituents being changed), little difference is found as one might expect. There are differences in the thermal decomposition properties, $\Delta\Delta E_{solv}$ and $\Delta\Delta S_{solv}$, however, these properties possess substantial errors and the differences are not believed to be significant; additional calculations at different λ values and/or longer calculations at $\lambda = 0.5$ are necessary for the convergence of these properties.

Interaction and binding thermodynamics

In Table 3 we list the results of our calculations on the relative thermodynamics of interaction for complexes of TMP, and two of its congeners, with the NADPH-DHFR complex.

Table 3 *Protein-drug interaction thermodynamics*[a]

System		ΔΔA (kcal/mol)	ΔΔE (kcal/mol)	ΔΔS (eu)
TMP → TET[b]	(a)	-4.1 ± 0.4	22.8 ± 13	90 ± 45
TMP → PET[b]	(a)	-2.8 ± 0.2	-11.7 ± 7	-31 ± 24
TMP → TMPH[+c]	(a)	-33.4 ± 0.3	-38.2 ± 9.0	-16 ± 30
TET → TETH[+c]	(a)	-32.3 ± 0.4	-37.0 ± 10.0	-16 ± 33
PET → PETH[+c]	(a)	-45.1 ± 0.2	-39.3 ± 6.0	19 ± 21

[a] All simulations were carried out using the stochastic boundary molecular dynamics method [21] with 1847 protein atoms, NADPH, the drug molecule and 32 TIP3P water molecules in a 14 Å reaction zone. Calculations were done at 298 ± 5 K using Langevin dynamics for temperature bath and a timestep of 1.5 fs.

[b] Double-wide sampling with $\lambda = (0.125, 0.5, 0.875, 0.969)$ using 15 ps thermalization and 21 ps production at each λ value.

[c] Double-wide sampling with $\lambda = (0.5, 1.0)$ using 7.5 ps thermalization and 30 ps production at each λ value.

With the results from Tables 2 and 3, we can construct the thermodynamics associated with a number of inhibitor exchange reactions; of greatest interest are the four indicated in Table 4.

Table 4 *Thermodynamics of important inhibitor exchange reactions*

Reactions	ΔΔA (kcal/mol)	ΔΔE (kcal/mol)	ΔΔS (eu)
(1) TET + DHFR:TMP ⇋ TMP + DHFR:TET	0.8 ± 0.5	26 ± 14	83 ± 50
(2) PET + DHFR:TMP ⇋ TMP + DHFR:PET	-1.3 ± 0.3	5 ± 9	20 ± 29
(3) TET + DHFR:TMPH$^+$ ⇋ TMP + DHFR:TETH$^+$	1.9 ± 0.7	27 ± 20	83 ± 67
(4) PET + DHFR:TMPH$^+$ ⇋ TMP + DHFR:PETH$^+$	-13.0 ± 0.5	4 ± 13	55 ± 47

For these reactions it is most noteworthy that the compound PET appears to bind more tightly than either TMP or TET. Furthermore, it is extremely interesting to notice the thermodynamic decomposition of the relatively small relative free energy changes associated with reactions (1)-(3); this small free energy change is a result of a large energy-entropy compensation - a result usually ignored in other calculations of relative binding thermodynamics. When the results for reaction (3) are compared with experiment [19], few structural differences are found for the TMP and TET complexes (see Fig. 5). However, the calculated thermodynamic value for the relative binding affinity, $\Delta\Delta A_{bind} = 1.8$ kcal/mol, differs from the experimental value of $\Delta\Delta G_{bind} = -1.7$ kcal/mol [19]. Given the large energy-entropy compensation observed above, this is not surprising since small errors in $\Delta\Delta E_{bind}$ or $\Delta\Delta S_{bind}$ can arise from a number of sources, e.g., potential parameters, sampling, etc. For the reaction involving protonated PET bound to the protein [reaction (4)], we predict a very large relative binding affinity compared to the other two compounds. This prediction will have to await further experiment for its verification.

Finally, we add a cautionary note in the interpretation of the present calculations, and in fact all thermodynamic calculations. This has to do with convergence and sampling statistics when the changes being considered may be accompanied by shifts in the relative positions of the reactant or product species. This point is illustrated in Fig. 5, which shows the positions taken by the three drugs in the protein binding pocket. A few of the key residues in the complexes are also included in the figure to orientate the viewer to the character of the interactions. What is clear in this figure is that the three drugs bind in somewhat different positions within the binding pocket. We believe that these alternate modes of binding may be attributed to the loss or disruption of attractive water-methoxy interactions between the solvent and the drug molecules, but the simple presence of different binding modes makes the calculation of relative thermodynamic changes difficult at best.

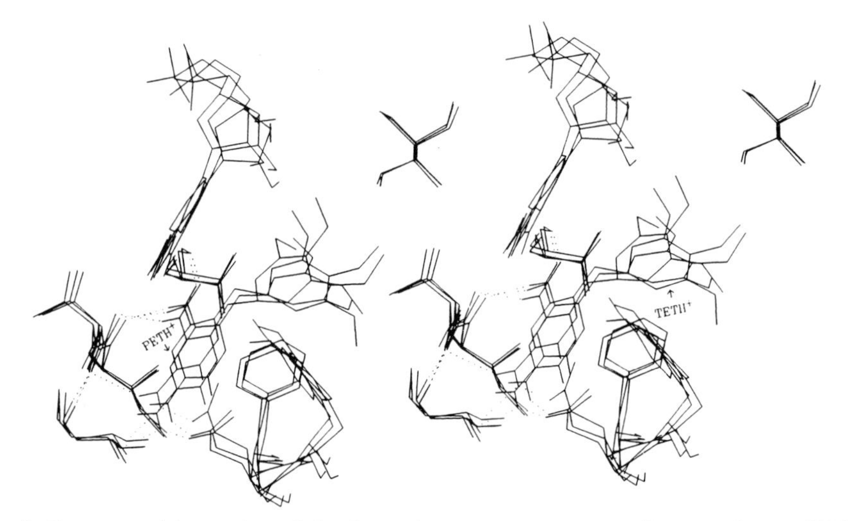

Fig. 5. Stereographic overlay of the dynamics average structures for complexes of DHFR with TMPH+, PETH+ and TETH+. Hydrogen bonds are indicated in the figure by dashed lines.

New Methods: Temperature Derivatives [3,7] and Free Energy Surfaces [12]

As indicated in the introductory and theory portions of this paper, we are also actively involved in developing new TS methods. We shall briefly discuss two new techniques being explored.

Temperature derivatives

In Table 5 we make a comparison of relative *energies* of solvation for a Lennard-Jones particle dissolved in Lennard-Jones 'argon', computed using the TS formula given in Eq. 6, with quantities computed using the conventional 'energy difference between two separate simulations'.

Table 5 *Thermodynamic derivative properties: energy and entropy*

System	ΔΔE (kcal/mol)		ΔΔS[a] (eu)
	Derivative formula[a]	Conventional method[b]	
Ar → Ar (2ϵ)[c]	−3.65 ± 0.3	−3.15 ± 1.0	−9.8 ± 3
Ar → Ar[d]	−1.07 ± 0.3	−1.1 ± 1.4	−10.2 ± 4

[a] Thermodynamic derivative connection formula as given in Eqs. 6 and 7 used in evaluation of ΔΔE and ΔΔS.

[b] Calculations consisted of two simulations of 30 ps duration production dynamics (first case) and 70 ps production dynamics (second case) each with a timestep of 2 fs.; ΔΔE computed from the difference in the average solute-solvent interaction energy.

[c] Simulations are for one solute atom ($\epsilon = 120$ K, $\sigma = 3.4$ Å) → (ϵ 240 K, σ 3.4 Å) and 107 solvent atoms (ϵ 120 K, σ 3.4 Å) at a density of $\rho = 0.02$ Å^{-3} and a temperature of 85 ± 3 K. Temperature was controlled with Langevin dynamics. Calculations used double-wide sampling for $\lambda = 0.5$ and consisted of 10 ps equilibration and 30 ps production dynamics.

[d] Simulations are for one solute atom ($\epsilon = 120$ K, $\sigma = 3.4$ Å) → $\epsilon = 132$ K, $\sigma = 3.74$ Å) and 107 solvent atoms ($\epsilon = 120$ K, $\sigma = 3.4$ Å) at a density of $\rho = 0.02$ Å^{-3}. Each calculation used $\lambda = (0.125, 0.5)$ with 10 ps equilibration and 70 ps production dynamics at each λ value.

As can be seen from results in this table, higher precision is achieved using the derivative formula. This is encouraging since many TS calculations are often carried out at λ values not equal to 1 or 0 and the overall thermodynamic cycle is constructed using the double-wide sampling scheme without sampling at the end points. Thus, energetic and entropic contributions may be deduced (with greater precision) without the necessity of doing additional calculations. Furthermore, as illustrated with our calculations on DHFR, the thermal decomposition of free energies provides additional thermodynamic data for use in the rationalization, prediction and design of new drug analogs with enhanced binding (or other) characteristics.

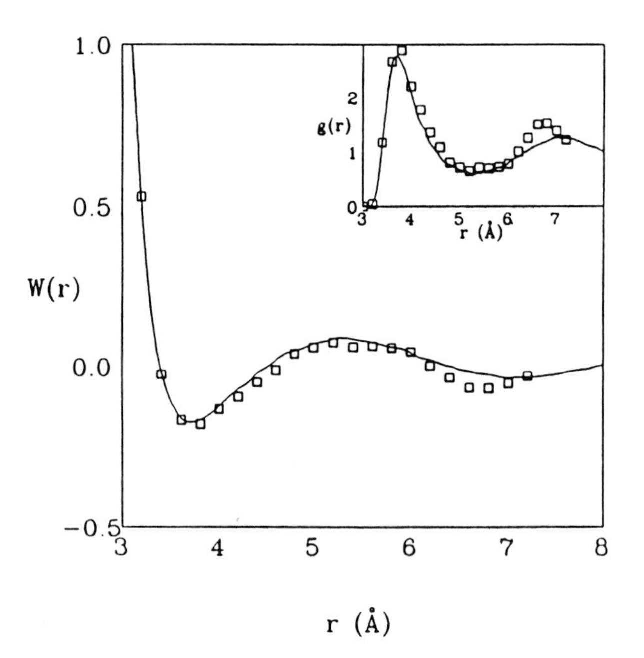

Fig. 6. The potential of mean force, W(r) from Eq. 10 in kcal/mol, and corresponding pair correlation function, g(r), for two labeled Ar atoms in 108 Ar atoms at 86 K (squares). The 'exact' results for computer simulation are shown as the solid line.

Free energy surfaces [12]

We have recently made the first application of TS methods using molecular dynamics to the calculation of free energy surfaces [12]. In this calculation we examined the feasibility of using this technique to explore reaction surfaces in a simple application to the potential of mean force for two argon atoms. Our results are encouraging, as is illustrated in Fig. 6.

We are able to reproduce the exact curve quite well for all atom-atom separations of interest. This methodology will be of great significance not only in the calculation of condensed phase reaction surfaces, but also as an additional sampling tool to be used in conjunction with the other TS methods discussed above in cases where conformational flexibility and alternate conformations are expected to be important in the overall thermodynamics of a given process. Such may be the case with the binding of TMP and its congeners. We are currently exploring this problem using the combined TS methodologies for free energies, energies, entropies and free energy surfaces.

Summary and Conclusions

At the present time there are a number of problems which need attention regarding TS methods and the underlying mathematical models and sampling techniques used in conjunction with them. We have pointed out possible inadequacies in the representation of short-range non-bonded interactions by empirical potential models; these need further investigation. We have also illustrated the additional information which may be obtained from TS calculations by computing the temperature-derivative properties; the future holds much promise for this technique. Finally, we have noted some of the problems associated with conformational sampling in thermodynamic calculations and provided a new approach to conquering this problem.

The overall outlook for these approaches is good. Thermodynamic simulation methods can be expected to stimulate and add to our knowledge of condensed phase (especially aqueous) processes. In particular, their application to problems of drug design and protein stability will continue and increase, and the fruit of these applications will multiply in its abundance as maturation of the techniques grows.

Acknowledgements

I would like to thank Dr. Stephen H. Fleischman for his contributions on the calculations presented for alkanes and alcohols in aqueous solution, for calculations on the DHFR:TMP systems and his assistance in the development of the TS methodologies. Mr. Douglas J. Tobias is acknowledged for his work on free energy surfaces. Finally, acknowledgement is made to the NIH (grant No. GM37554-02) for financial assistance and to the Pittsburgh Supercomputer Center (grant Nos. PSC34, PSC34A and PSCA282) for computer time on the CRAY X-MP.

References

1. Brooks III, C.L., Karplus, M. and Pettitt, B.M., Proteins: A Theoretical Perspective on Structure, Dynamics and Thermodynamics, In Advances Chemical Physiology, Vol. 71, 1987.
2. Mezei, M. and Beverige, D.L., Ann. N.Y. Acad. Sci., 483 (1983) 1.
3. Fleischman, S.H. and Brooks III, C.L., J. Chem. Phys., 87 (1987) 3029.
4. Jorgensen, W.L. and Ravimohan, C., J. Chem. Phys., 83 (1985) 3050.
5. (a) Bash, P.A., Singh, U.C., Brown, F.K., Langridge, R. and Kollman, P.A., Science, 235 (1987) 574.
   (b) Singh, U.C., Brown, F.K., Bash, P.A. and Kollman, P.A., J. Am. Chem. Soc., 109 (1987) 1607.

6. (a) Wong, C.F. and McCammon, J.A., J. Am. Chem. Soc., 108 (1986) 3830.
   (b) Lybrand, T.P., McCammon, J.A. and Wipff, G., Proc. Natl. Acad. Sci. U.S.A., 83 (1986) 833.
7. Fleischman, S.H. and Brooks III, C.L., Science, submitted for publication.
8. Fleischman, S.H., Tidor, B., Brooks III, C.L. and Karplus, M., J. Comput. Chem., manuscript in preparation.
9. Zwanzig, R.W., J. Chem. Phys., 22 (1954) 1420.
10. Tembe, B.L. and McCammon, J.A., Computers & Chemistry, 8 (1984) 281.
11. (a) Brooks III, C.L., J. Phys. Chem., 90 (1986) 6680.
    (b) There were typographical errors in Eqs. 8 and 9 of the above reference, the corrected relationships are given in Eqs. 6 and 7 of the present paper.
12. Tobias, D.T. and Brooks III, C.L., Chem. Phys. Lett., 142 (1987) 472.
13. (a) Ben-Naim, A. and Marcus, Y., J. Chem. Phys., 81 (1984) 2016.
    (b) Dec, S.F. and Gill, S.J., J. Solution Chem., 13 (1984) 27.
14. Brooks, B.R., Bruccoleri, R.E., Olafson, B.D., States, D.J., Swaminathan, S. and Karplus, M., J. Comput. Chem., 4 (1983) 187.
15. Smith, J. and Karplus, M., J. Comput. Chem., to be submitted.
16. Jorgensen, W.L., Madura, J.D. and Swenson, C.J., J. Am. Chem. Soc., 106 (1984) 6638.
17. Mendenhall, W., Introduction to Probability and Statistics, Duxbury Press, North Scituate, MA, 1975.
18. Jorgensen, W.L., Binning Jr., R.C. and Bigot, B., J. Am. Chem. Soc., 103 (1981) 4393.
19. (a) Matthews, D.A., Bolin, J.T., Burridge, J.M., Filman, D.J., Volz, K.W., Kaufman, B.T., Beddell, C.R., Champness, J.M., Stammers, D.K. and Kraut, J., J. Biol. Chem., 280 (1985) 381.
    (b) Matthews, D.A., Bolin, J.T., Burridge, J.M., Filman, D.J., Volz, K.W. and Kraut, J., J. Biol. Chem., 280 (1985) 392.
20. Jorgensen, W.L., Chandrasekhar, J., Madura, J.D., Impey, R.W. and Klein, M.L., J. Chem. Phys., 79 (1983) 926.
21. (a) Brooks III, C.L., Brunger, A.T. and Karplus, M., Biopolymers, 24 (1985) 434.
    (b) Brooks III, C.L. and Karplus, M., Methods Enzymol., 127 (1986) 369.
    (c) Brooks III, C.L. and Karplus, M., J. Mol. Biol., in press.

Free energy perturbation calculations in drug design applications

Terry P. Lybrand
University of Minnesota, Department of Medicinal Chemistry, Health Sciences Unit F, Minneapolis, MN 55455, U.S.A.

Various computer modeling techniques have been used for some time now to aid in the drug design process [1-6]. In cases where good three-dimensional structural data is available for both drug and target receptor site, computer modeling techniques such as interactive computer graphics and molecular mechanics calculations can provide important information about the nature of drug-receptor interactions. This information can then be used to guide the development of new drug molecules that may have improved properties. A recently developed computational technique, the thermodynamic cycle-perturbation method, should prove to be an especially powerful tool in the computer-assisted drug design process [7, 8]. This method allows the quantitative calculation of relative binding free energies for a series of related drug molecules at a common receptor site and has been used with good success in several applications projects [9-11]. With this technique, it is now possible to predict how substituent modification in a lead compound will affect binding affinity at the receptor site.

Usually, three-dimensional structural data for drug and receptor molecules is obtained from X-ray crystallography studies (e.g. crystal structures for an enzyme-inhibitor complex). Unfortunately, good crystal structures are available for only a limited number of interesting target receptor sites. However, it is possible to construct reasonable three-dimensional structures for many drug-receptor complexes using computer modeling techniques and experimental data. The example discussed below will outline a general method for constructing model structures of an actinomycin D-DNA complex, as well as the application of the thermodynamic cycle-perturbation method and other computational techniques in an effort to 'design' sequence-selective DNA binding molecules.

Actinomycin D (AMD, Fig. 1) is a chromopeptide antibiotic with limited therapeutic utility in the treatment of certain tumors [12]. Extensive experimental studies have shown that the AMD chromophore intercalates selectively on the 3′ side of guanine residues in double-stranded DNA, with the cyclic pentapeptide side chains lying in the minor groove of the DNA helix [13-16]. Crystal structures of AMD with deoxyguanosine [17] and short nucleotides [18, 19] are also

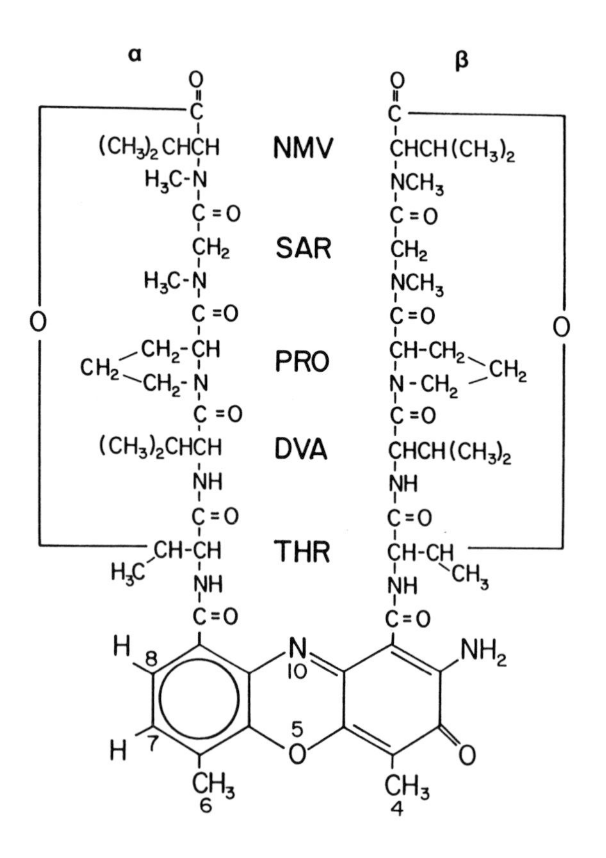

Fig. 1. Actinomycin D.

supportive of these general details. Using this experimental data, it was possible to construct three-dimensional structures of AMD complexes with various double-stranded hexanucleotide complexes. The model structures were created using interactive computer graphics and energy minimization techniques. Finally, the model structures were compared with results from two-dimensional NMR-NOE experiments [20]. Of 214 proton-proton distances compared, only 12 (6%) displayed large deviations (≥ 1.5 Å) between model structures and NOE data [4]. A molecular dynamics simulation and analysis for the model structures further improves the 'agreement' with NOE data [21]. Thus, the model-built AMD-hexanucleotide complexes appear to be reasonable representations for the solution phase structures of these complexes. Of course, it is possible to incorporate the two-dimensional NMR data, or other spectroscopic data, as constraints in the model building process, if sufficient data is available when the modeling studies are initiated.

Initial studies of the AMD-hexanucleotide model complexes utilized computer graphics and energy minimization methods to analyze details of the drug-receptor interactions [4]. Preliminary results indicated that it might be possible to introduce sequence-selective binding properties in AMD through modification of the amino acid side chains, since a graphical analysis of the complexes showed that the pentapeptide side chains have extensive interactions with the DNA bases in the floor of the minor groove. Gas phase energy minimization results suggested that AMD in fact prefers to bind the sequence $(ATGCAT)_2$ as compared to $(GCGCGC)_2$. Graphical and numerical analysis of results suggested that this apparent binding preference was due to unfavorable steric interactions between the *N*-methyl group of *N*-methyl valine (NMV) in AMD and the exocyclic amino group of the 5′-terminal guanines in $(GCGCGC)_2$. While intriguing, these results are only qualitative at best, due to the limitations of the calculations. In order to get more reliable information from the computer models, it is necessary to (1) include solvent and counterions explicitly in the calculations, and (2) compute binding free energies rather than internal energies (as in energy minimization calculations). The rapid escalation in computer power over the past several years now makes the first enhancement feasible, and new techniques like the thermodynamic cycle-perturbation method allow direct calculation of relative binding free energies for different complexes.

Calculations are under way now, using the thermodynamic cycle-perturbation method, to determine if AMD does bind preferentially to the $(ATGCAT)_2$ sequence. These calculations involve the perturbation of a terminal G-C base pair to an A-T base pair in the hexanucleotide to compute the sequence binding preference. The results should provide a much more reliable assessment of AMD base sequence preference than previous calculations, as well as a much more detailed picture of the intermolecular interactions that determine any binding preferences. These calculations will also provide important information about the relative ease of introducing sequence-selective binding properties. If the presence (or absence) of a methyl group can select for an A-T versus a G-C base pair, there is reason to hope that other selectivity can be introduced in AMD derivatives without too much difficulty. Subsequent calculations will focus on pentapeptide side chain substituent modifications that may impart some sequence-selective binding properties. Later, these studies will be extended to include other molecules that exhibit DNA base sequence selectivities that are quite different from AMD. Using the information obtained in these calculations, it should be possible to compile a 'data base' of molecular fragments (components of AMD and other molecules) with well defined base sequence binding preferences. This data base may then provide useful templates for the de novo design of sequence-selective DNA binding molecules.

Although free energy perturbation techniques have been used with great success, there are still cases where some degree of caution is warranted. Relative free energies of hydration for hydrophobic species can be especially difficult to compute accurately (C. Brooks, B.M. Pettitt, personal communications). It is also difficult to ensure that free energy calculations have 'converged', especially when torsional degrees of freedom are intimately affected by the substituent perturbations (T. Lybrand, unpublished results). It should be possible to overcome these computational limitations by modification of perturbation algorithms or potential energy functions, or both.

Calculations that were unthinkable a few years ago for many biochemical systems are now feasible, thanks to the development of new computational techniques like the free energy perturbation methods and advancements in computer hardware capability. The ability to perform such calculations holds great promise in drug design and protein engineering applications.

Acknowledgements

T.P.L. is the recipient of a National Science Foundation Presidential Young Investigator Award (DMB-8657042) and a University of Minnesota McKnight-Land Grant Professorship. This work has also been supported in part by a Damon Runyon-Walter Winchell Cancer Fund fellowship (DRG-888).

References

1. Hansch, C., Li, R., Blaney, J.M. and Langridge, R., J. Med. Chem., 25 (1982) 777.
2. Wipff, G., Blaney, J., Weiner, P., Dearing, A. and Kollman, P.A., J. Am. Chem. Soc., 105 (1983) 997.
3. Froimowitz, M., Salva, P., Hite, G.J., Gianutsos, G., Suzdak, P. and Heyman, R., J. Comput. Chem., 5 (1984) 291.
4. Lybrand, T.P., Brown, S.C., Creighton, S., Shafer, R.H. and Kollman, P.A., J. Mol. Biol., 191 (1986) 495.
5. Wong, C.F. and McCammon, J.A., J. Am. Chem. Soc., 108 (1986) 3830.
6. Tonani, R., Dunbar Jr., J., Edmonston, B. and Marshall, G.R., J. Comput.-Aided Mol. Design, 1 (1987) 117.
7. Lybrand, T.P., Ghosh, I. and McCammon, J.A., J. Am. Chem. Soc., 107 (1985) 7793.
8. Lybrand, T.P., McCammon, J.A. and Wipff, G., Proc. Natl. Acad. Sci. U.S.A., 83 (1986) 833.
9. Lybrand, T.P., Lau, W.F., McCammon, J.A. and Pettitt, B.M., In Oxendar, D. (Ed.) Protein Structure, Folding, and Design. UCLA Symposia on Molecular and Cellular Biology, Vol. 69, Alan R. Liss, Inc., New York, 1987, p. 227.
10. Bash, P.A., Singh, U.C., Langridge, R. and Kollman, P.A., Science, 236 (1987) 564.
11. Van Gunsteren, W.F. and Berendsen, H.J.C., The power of dynamic modelling of molecular systems, to be submitted to J. Comput.-Aided Mol. Design.
12. Perry, S., Cancer Chemother. Rep., 58 (1974) 117.
13. Müller, W. and Crothers, D., J. Mol. Biol., 35 (1968) 251.
14. Patel, D., Biochemistry, 13 (1974) 2388, 2396.

15. Reinhardt, C.G. and Krugh, T.R., Biochemistry, 16 (1977) 2890.
16. Sengupta, S.K., Trites, D.H., Madhavarao, M.S. and Beltz, W.R., J. Med. Chem., 22 (1979) 797.
17. Sobell, H.M., Jain, S.C., Sakore, T.D. and Nordinan, C.E., Nature New Biol., 231 (1971) 200.
18. Takusagawa, F., Dabrow, M., Neidle, S. and Berman, H., Nature, 296 (1982) 466.
19. Takusagawa, F., Goldstein, B.M., Youngster, S., Jones R.A. and Berman, H.M., J. Biol. Chem., 259 (1984) 4714.
20. Brown, S.C., Mullis, K., Levenson, C. and Shafer, R.H., Biochemistry, 23 (1984) 403.
21. Creighton, S., Rudolph, B., Lybrand, T., Singh, U.C., Shafer, R., Brown, S., Kollman, P., Case, D. and Andrea, T., J. Biomol. Struct. Dyn., submitted.

Successes, failures and curiosities in free energy calculations

B. Montgomery Pettitt
Department of Chemistry, University of Houston, Houston, TX 77004, U.S.A.

A number of different theoretical and computational approaches to the calculation of the free energy difference between different chemical or thermodynamic states have recently been used on systems of increasing complexity [1, 2]. The change in chemical potential, $\Delta\mu$, or the calculation of a free energy change from a reference system are sensitive measures of the quality of both the method and the models employed. The relative change in free energy between systems within a given thermodynamic cycle, $\Delta\Delta A$, can potentially be a more reliable number since the characteristics of the reference system tend to cancel out.

In this contribution the sensitivity of the free energy difference in a variety of systems will be explored. The calculations tend to show a dependence on both methods and models. Among the various simulation methods are the particle insertion technique, the windowing or bias technique, the slow growth methods, all of which may be implemented with either Monte Carlo or molecular dynamics configurational sampling [1, 2]. In addition, there exist the various analytic methods [1]. Many of these reduce to the numerical integration of a one-dimensional function; in some cases an exact differential of the free energy may be found and integrated at once [3]. A number of model forms for all of the above-mentioned calculational techniques are in common use. These include the extended-atom representation where all non-hydrogen bonding hydrogens are included in the nearest heavy atom, and the so-called all-atom representation. Each of these types of models have different parameterizations and may include nonadditive, nonbonded interactions such as either molecule centered or atom-based polarizabilities.

This diversity of methods and models can be applied both to intermolecular problems, such as binding constants or partition coefficients, and to intramolecular problems such as conformationally dependent free energies. A number of relevant studies fall into the category of a combination of these two types, such as reactions and the modeling of transition state analogs where intermolecular and intramolecular effects are intertwined. The examples to be given to illustrate this area will range from 60 Da to 10^7 Da.

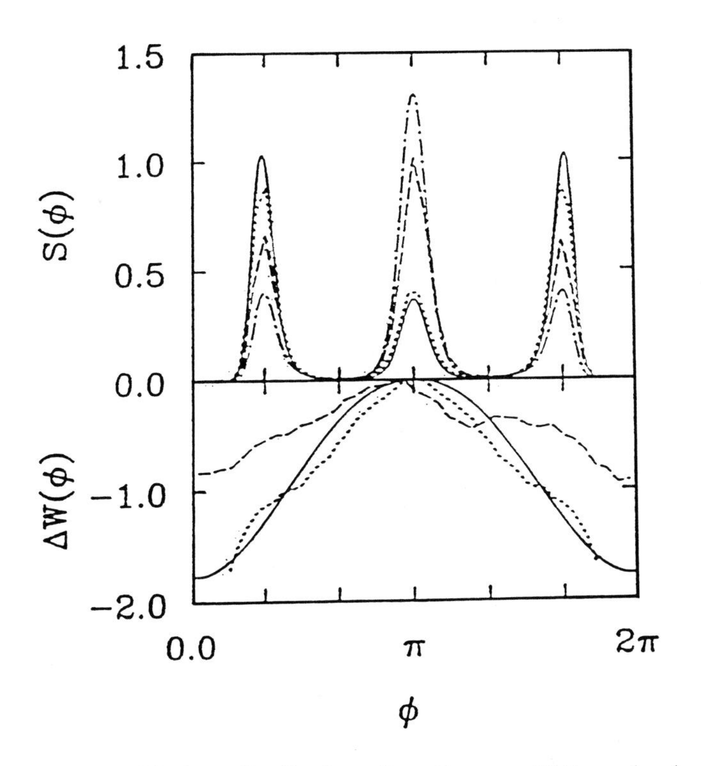

Fig. 1. Normalized equilibrium distribution of conformers, $S(\phi)$, and solvent-induced dihedral free energy, $\Delta W(\phi)$ in kcal/mol for n-butane in H_2O; ideal gas phase (- · -); simulation data of Jorgensen [4], (- - -); simulation data of Rosenberg et al. [5], (· · ·); XRISM (—).

The reason that we must use all of this free energy machinery to make adequate predictions of chemical interest is that simple energy calculations are misleading for systems which are neither in vacuo nor at absolute zero in temperature. Two fundamental questions might be asked at this point. The first is: Why do effective inter- and intramolecular potentials change on going from gas phase to liquid (aqueous) phase? The two major reasons are:

(1) Packing effects, which are due to excluded volume (molecular shape) in the manybody fluid bath.

(2) Dielectric screening effects, which are due to manybody correlations in polar and polarizable fluid baths.

The second question is: How can one understand the macroscopic properties in terms of microscopic molecular models and properties? Clearly, at least two of the solutions are:

(1) Computer simulations, which can be Hamiltonian, Monte Carlo or Langevin.
(2) Analytic methods, such as integral equations and field theories.

Examples

The first example that will be considered here is that of the conformational free energy differences in *n*-butane about the central carbon-carbon torsion. The exponential of the coordinate or species-dependent free energy is proportional to the population at the particular value of the coordinates or species. Thus we may choose to consider populations, free energy changes or ratios of populations, in the form of equilibrium constants without loss of generality. In Fig. 1 the Boltzmann population as a function of the torsion angle, $S(\phi)$, is given for gas phase and a number of aqueous phase estimates.

Clearly the various theoretical estimates all agree in the direction of the so-called 'hydrophobic shift' in the population on going from gas to aqueous phase; namely, that the trans conformer is depopulated relative to the gauche conformers. However, it is equally clear that the theoretical estimates are hardly in accord about the magnitude of the population shift. Curiously, the two methods that one might expect to be the closest, namely the two computer simulations, essentially span the range of the disagreement with the other analytically based techniques generally giving estimates less extreme. Besides the difference in methods, there are also differences in the models used and it is not clear at present where the blame should lay. With this sort of accuracy in mind, that is, that the direction is probably reasonable, if not the actual value of the equilibrium constant for changing between rotamers, we can proceed to consider more complicated molecules.

The next molecule to be considered here is *N*-acetyl alanine *N'*-methyl amide or the alanine dipeptide (Fig. 2). This system has the two classic Ramachandran dihedral angles, ϕ,ψ, and contains two complete peptide moieties. That, in addition to a rudimentary side chain and a large amount of previous experimental (and theoretical) work, makes this an interesting test for methods to be used on molecules of biophysical interest. In fact, this dipeptide and several of its analogs have recently been shown to form a class of potent antiepileptics [6].

The intramolecular potential surface for these two degrees of freedom has been calculated by a number of different quantum chemical methods. Most such calculations agree that if the molecule is allowed to relax adiabatically while the mapping for the ϕ,ψ combinations is proceeding, two stable intramolecularly hydrogen-bonded conformers are found which produce pseudo seven-membered rings with the β-methyl group in either the axial or equatorial position relative to the hydrogen-bonded ring. These are designated by C_{7ax} and C_{7eq}, respectively. Unfortunately, given a decent quantum-mechanical surface, there is no unique

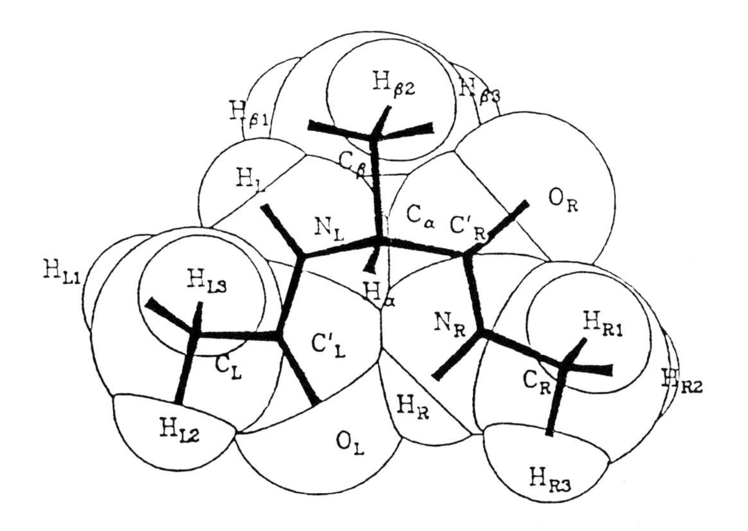

Fig. 2. Space filling model of the alanine dipeptide.

decomposition into the terms of a classical model which generally includes the familiar bond stretch, angle stretch, dihedral rotation, improper dihedral distortion and nonbonded interaction terms. Because solvent effects are different for each of these interactions, different parameterizations can yield substantially different results.

Older force fields frequently used an inverse 10th power attraction for hydrogen bonds [7]. More recent work tends to model the hydrogen bond as an electrostatic effect which thereby uses an inverse first power interatomic radial dependence. Each of these models can be fit to the quantum mechanical data within a satisfactory tolerance. Thus, in vacuo (dilute gas phase), these models give quite similar answers for properties such as the conformational Boltzmann population even though they have different charge distributions and therefore different dipole moments. The gas phase ϕ, ψ populations for two such models are shown in Fig. 3.

The liquid state dihedral population distributions are strongly affected by the charge distribution of both the solute molecules and the solvent molecules. This is a reasonable consequence of the screening of charge-charge interactions in a dipolar fluid yet both models fit the total gas phase quantum mechanical surface. Thus it is important to include as much extra data, such as dipole moment data, as possible in the parameterization. Another way of looking at the cause for the population differences between these models in solvent, displayed in Fig. 4, is that the solute model had a potential intramolecular hydrogen bond

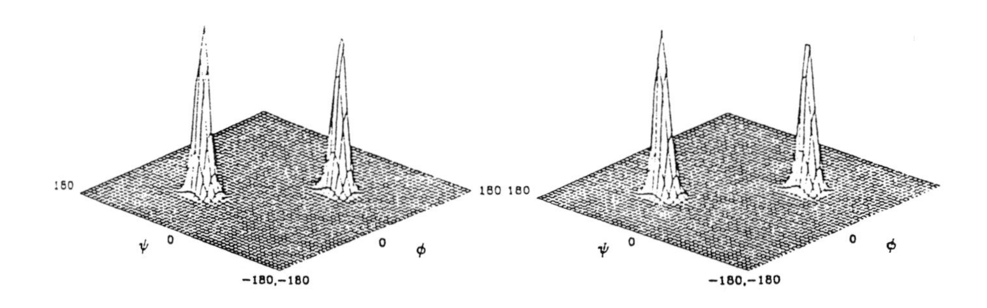

Fig. 3. The gas phase φ,ψ populations for two model systems.

that could not be seen by the solvent because the aqueous solvent model was using a different force law for the hydrogen bonds (pure Coulomb) [1].

The final example to be used in this discussion will be a problem involving the interaction of drugs with a common cold virus, HRV-14. The cold virus is a member of the picornavirus family. This group of viruses consists of a spherical protein coat filled with RNA and solvent. The picornavirus assembly is roughly 300 Å diameter which is small compared to most other viruses or bacteria. Antiviral compounds of the oxazole family (isoxazole-heptane-phenoxy-oxazole) have been developed by the Sterling Winthrop Research Institute. These have been shown to have limited activity against the picornavirus family, of which cold viruses are members.

To begin our work, we required atomic coordinates of all the atoms in the system. For the theoretical studies discussed below we used the data from the X-ray crystal structure of the virus with the drug taken at the Cornell synchrotron provided by Michael Rossmann and coworkers from Purdue University. The point of the study was to determine the binding affinities of a number of different analogues in the Winthrop-oxazole family. Specifically, we first concentrated on the H → Cl replacement ortho to the oxygen of the phenoxy oxygen on the phenyl ring. The free energies were calculated with the time-dependent replacement method sometimes referred to as the 'slow growth' technique. Using a wide variety of starting conditions and functions of the replacement coordinate versus time, we found the total final answers were frequently in better accord than the answers at any time in the run other than the endpoints. Since any particular time *could* have been the physical endpoint of the computer experiment we must consider the errors to at least the greatest of the differences found along the replacement coordinate at any time. Thus, rather than the rms difference at the endpoint of ±0.1 kcal/mol we believe our errors to be closer to ±1.0 kcal/mol. Interestingly, we found that the free energy decomposition in terms of its van der Waals and electrostatic components along the replacement

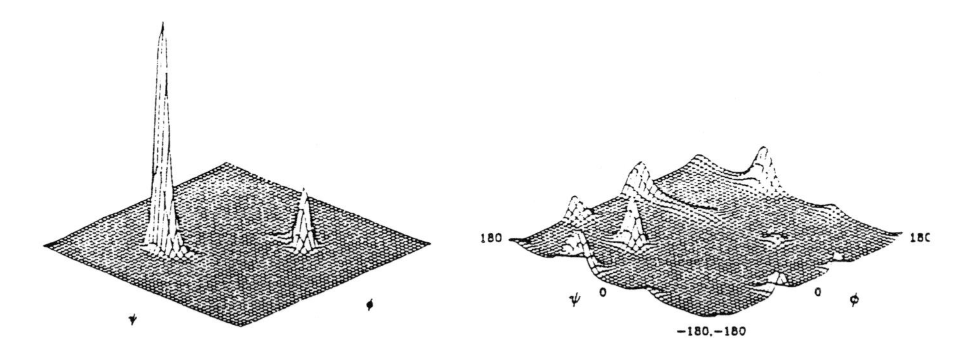

Fig. 4. The solvent phase φ,ψ populations for two model systems.

coordinate differed even more substantially. For runs with the same initial conditions, both 40-50 ps in length but one using a sinusoidal variation and one a linear variation in the replacement coordinate with respect to time, we found that the signs of the van der Waals and electrostatic contributions had been interchanged and balanced in such a way that the same final total free energy change resulted. With this in mind we now estimate that the errors are more like the sum of the errors in the energy components yielding an rms of ±3.0 kcal/mol. Thus the fact we obtained -0.5±3.0 kcal/mol is not in disagreement with the experimental number of +1.5 kcal/mol; it is simply not very satisfying.

A number of other common assumptions are frequently made in such binding studies. An addition replacement of Me→H was made at the first carbon on the oxazole ring. In that case, the experimental data appeared to agree with the result from simulation in both magnitude and sign. Unfortunately, after the work was finished (but before it was published) we discovered that recent crystallographic evidence found that the binding geometry for the methyl and *des*-methyl compounds was substantially different. In fact, the molecule had performed a 180° flip in the pocket in the long dimension. Thus, the assumption of a similar binding geometry among homologs can be quite serious.

While free energy techniques hold a great deal of promise for the future, it is clear that a great deal more careful work must be performed to establish their quantitative range of validity. The future of these techniques for considering both intermolecular and intramolecular changes will depend on our ability to find the reasons behind both the successes and the failures of the methods.

Acknowledgements

The author would like to thank his coworkers, Wan Fang Lau and Liem Dang, without whose hard work scarcely anything would have been done. In addition, M. Karplus, P. Rossky, T. Lybrand, A. McCammon, M. Rossman and M. McKinlay are acknowledged as collaborators on different aspects of the work discussed. Computer time was provided primarily through several grants at the NSF Supercomputer Center at San Diego. Partial financial support was provided by the R.A. Welch foundation and the State of Texas through the TATR program.

References

1. Brooks, C.L., Karplus, M. and Pettitt, B.M., Adv. Chem. Phys. (1988) in press.
2. McCammon, J.A. and Harvey, S., Dynamics of Proteins and Nucleic Acids, Cambridge University Press, Cambridge, 1987.
3. Singer, S. and Chandler, D., Mol. Phys., 55 (1985) 621.
4. Jorgensen, W.L., J.Chem. Phys., 77 (1982) 5757.
5. Rosenberg, R.O., Mikkilineni, R. and Berne, B.J., J. Am. Chem. Soc., 104 (1982) 7647.
6. Conley, J.D. and Kohn, H., J. Med. Chem., 30 (1987) 567.
7. See for instance: Rossky, P.J., Karplus, M. and Rahman, A., Biopolymers, 18 (1979) 825 and references therein.

Free energy perturbation calculations: Problems and pitfalls along the gilded road

David A. Pearlman and Peter A. Kollman
Department of Pharmaceutical Chemistry, University of California at San Francisco, San Francisco, CA 94143-0446, U.S.A.

Summary

The free energy perturbation (FEP) method for calculating the free energy difference between two states is one of great promise. But there are still many pitfalls surrounding and questions to be answered regarding these calculations. We describe and illustrate a number of these types of issues, concluding that until the method is better characterized, the results of FEP calculations, particularly on complex systems, should be carefully scrutinized.

Introduction

In a general sense, the Holy Grail of theoretical chemistry can be considered a method meeting four core criteria: (i) it has a sound scientific basis; (ii) it is applicable to both large and small molecular systems; (iii) it produces quantitative results with direct relevance to important experimentally measurable quantities; and (iv) it has a feasible and reliable implementation for all molecules. So strong is the lure of such a method, that even the possibility of Grail-dom being associated with a technique is sufficient to spawn an explosion of scientific endeavors using that technique. Such has been the case with so-called free energy perturbation (FEP) calculations [1]. These are the current darlings of the theoretical literature. Unfortunately, as will be discussed below, while a few trial FEP calculations have demonstrated tremendous potential for the method [2], the current attraction of these calculations is still based as much on promise as on demonstrated performance. Yet, because these calculations sport certain irresistibly-appealing features, their application is being made to some systems prematurely, and in some cases, unreliably.

FEP calculations are the latest step in a long heritage of theoretical calculations. In a sense, the grandfather of the family is the field of ab initio quantum calculations. Considered solely from the vantage point of scientific rigor, these are the most satisfying of theoretical methods, since the Schrödinger equation they attempt to solve contains an exact description of all properties of a molecular species. Without a doubt, these calculations satisfy core criteria i and iii, but they completely fail core criterion ii. They are inordinately computer-time expensive and a near-exact solution for nearly any polyatomic system is currently impossible [3]. Simplifying methods falling under the umbrella of quantum methods have been developed [4, 5], allowing systems of a few dozen or so atoms to be considered, but because they incorporate simplifications of unproven quality or uncertain effect, these methods ultimately fail criterion iv when applied to systems of this size: reliability of results. In addition, quantum mechanics must often be combined with statistical mechanics to extract experimentally relevant quantities. And quantum methods still cannot be directly applied to either macromolecular systems or systems including explicit solvent.

In an attempt to circumvent the inability to apply quantum methods to large molecular systems, empirically-based methods have been developed over the past two and a half decades [6-9]. These methods are based on the use of an empirical classical potential energy function to calculate the energy corresponding to any conformation of a system. A commonly-used energy function [9] takes the form

$$V_{Total} = \sum_{bonds} K_r (r - r_{eq})^2 + \sum_{angles} K_\theta (\theta - \theta_{eq})^2 + \sum_{dihedrals} \frac{V_n}{2} [1 + \cos(n\phi - \gamma)] +$$

$$\sum_{i<j} \left\{ \epsilon_{ij} \left[\left(\frac{R^*_{ij}}{R_{ij}} \right)^{12} - 2 \left(\frac{R^*_{ij}}{R_{ij}} \right)^6 \right] + \frac{q_i q_j}{\epsilon R_{ij}} \right\} \quad (1)$$

In this equation V_{Total} is the potential energy of the system; K_r and r_{eq} are the bond stretching constant and the equilibrium bond distance; K_θ and θ_{eq} are the bond angle stretching constant and the equilibrium bond angle; V_n, n, and γ are the torsional force constant, the periodicity of the torsional term, and the phase angle; ϵ_{ij} and R^*_{ij} are the non-bond (Lennard-Jones) well-depth and equilibrium interaction distance between particles i and j; R_{ij} is the interatomic distance between atoms i and j; q_i and q_j are the atomic partial charges on atoms i and j; and ϵ is the effective dielectric constant. The analytic forms and coefficients used in this expression have been derived and optimized to reproduce a selection of important experimental data [9, 10].

Clearly, methods based on such an energy function sacrifice to some degree core criterion i (scientific rigor), to gain in the area of criterion ii - applicability to interesting systems. This is acceptable as long as (1) the scientific rigor is still sufficient, i.e., the potential energy function is accurate enough to reliably predict the energy properties we are interested in; and (2) the other core criteria (iii and iv) can also be met in methods built around the energy function. Most of the remaining discussion will assume (1) [11], and consider (2) in more depth.

Once the analytic potential energy function had been introduced and parameterized, the first widespread application of this equation was energy minimization calculations. In these calculations, the equation

$$\frac{\partial V(\mathbf{r})}{\partial \mathbf{r}} = 0 \tag{2}$$

where **r** is the set of coordinates for a system, is solved iteratively. Implementation is easy and reliable (criterion iv), but the results are in many cases of only indirect relevance to experiment. This is both because of the 'local minimum problem' [12] - we can only find minima close to a starting guess (and not necessarily the most stable conformation) - and because such a method does not take into account or yield any dynamic or entropic information about a system. Both of the latter types of information are critically important in determining the experimental properties of a molecule. Conformational searching approaches can be used to partially circumvent the local minimum problem [13], but still fail to yield dynamic/entropic data.

A little over a decade ago the method of molecular dynamics (MD) was introduced to the study of macromolecular systems [14]. An integration of Newton's equation of motion,

$$\mathbf{F} = -\nabla V(\mathbf{r}) = m\mathbf{a} = m\frac{d^2\mathbf{r}}{dt^2} \tag{3}$$

MD promised to finally allow the calculation of dynamical molecular properties - properties that would be, in principle, relatable to experimental phenomena. Unfortunately, there are two large problems with MD. First, the complexity of Eq. 3 means that it can only be integrated using approximate methods. Such a solution severely limits the integration time-step that can be used (and so, because each step requires a costly computer evaluation of molecular forces, the time of the total trajectory that can be simulated). Second, while MD simulations generate a huge amount of data, it has been unclear what to do with the data once it is generated. Bulk properties such as diffusion constants and radial distribution functions can be calculated from MD trajectories of pure solvents. But procedures for analysis of this type of data on more complex (and interesting) systems in a manner that reveals experimentally relevant and reliable information have not been forthcoming.

At least not until a few years ago. At that time, several labs [1, 15-19] finally initiated development of a procedure that uses MD, not to generate time-dependent trajectory data, but rather as a means of generating an ensemble of conformational states that can be used to evaluate statistical mechanical quantities. From equations of elementary classical statistical mechanics, it can be straightforwardly derived that

$$\Delta G = G_b - G_a = -RT \ln \langle e^{-\Delta V/RT} \rangle_a \qquad (4)$$

where $\Delta V = V_b - V_a$ is the difference between the potential energy functions describing states 'b' and 'a', both evaluated for the same set of coordinates. $\langle\rangle_a$ means the ensemble average of the quantity within the brackets is to be calculated, by generating an equilibrium ensemble of states using the potential energy function corresponding to state 'a' and evaluating the quantity within the brackets for each of them. R is the molar gas constant, and T is the temperature. (A momentum contribution to ΔG is generally considered negligible. It can be calculated analytically, if necessary.)

What Eq. 4 says is that from an ensemble average of a potential energy difference (which we can evaluate using MD), we can derive the free energy difference between two states. This equation is not new [20]. But the technology allowing the requisite ensemble average to be evaluated using molecular dynamics did not exist until the advent and proliferation of high-speed computers.

As we have discussed, core criteria i and ii are adequately satisfied by this method (an application of MD, which is based on the potential energy function, and can be performed on both small and large systems). And since the free energy is the single most important factor regulating most experimental processes, core criterion iii is fulfilled as well. But what of core criterion iv? Specifically, can Eq. 4 be implemented reliably? The remainder of this paper will be devoted to exploring this question.

The simplicity of Eq. 4 is both exciting and misleading. Because while it is true that an exact evaluation of the ensemble average of the quantity in brackets will exactly yield the free energy difference, for any complex system it is impossible to determine, with absolute certainty, whether the ensemble average has converged.

And this indeterminacy of convergence is the underlying problem with free energy perturbation calculations.

For any reasonably complex potential surface, we can never know absolutely that we have sampled all accessible conformations that would contribute significantly to the ensemble average. We have no *sufficient* condition for convergence. However, we can characterize *necessary* conditions for convergence. Additionally, we can develop simulation methodologies which may allow convergence to be approached more rapidly. And for small model systems, we can use accurately known experimental free energies as stringent criteria on the convergence of these calculations.

Only after the development of a reliable and consistent set of procedures for carrying out FEP calculations on such well-characterized test systems has been accomplished can we confidently apply FEP methods to more elaborate and interesting systems. Unfortunately, many possible pitfalls which accompany these methods are easily overlooked, particularly because a single (free energy) number results from a calculation, and regardless of the accuracy of the calculation, this number can often agree at least qualitatively with expectation. But without sufficient checks, agreement with expectation in any particular case may be fortuitous.

Below we outline several potential problems, and a few methodological issues that can be encountered when performing free energy calculations. Several of the results are parts of 'work in progress', and the discussion is not intended to be all-inclusive. Nor do we claim to have answered all (or even most) of the questions we raise. Significantly more testing and characterization of these types of calculations will be required before a good set of answers can be determined. But we hope that this discussion will at least remind the user of FEP methods that it is still necessary to be very careful in applying these calculations. These methods have soaring promise. It is up to us as critical and careful scientists to ensure we make good on that promise.

Methods

The background equations for performing standard MD and FEP calculations have been discussed in detail elsewhere [1, 19, 21]. Here we simply outline the most important points.

Equation 4 can be considered the 'master equation' of FEP calculations. As we have noted, the main problem when implementing this equation is accurately calculating the ensemble average of the quantity within the brackets. We must calculate the difference between the potential energies of the system for two different potential functions (those corresponding to states 'a' and 'b'), but using the same set of coordinates - those from the trajectory generated using the potential corresponding to state 'a'. For this reason, if states 'a' and 'b' are too disparate, we will never sample important low-energy conformations of state 'b'. An extreme (but often encountered) example is one where state 'a' corresponds to a solvated particle and state 'b' corresponds to only the solvent. In this case, certain low-energy conformations of state 'b', the pure solvent, will never be sampled because in the potential used to generate the conformations, there is a particle creating a hole in the solvent [15]. In less extreme examples, sampling is still a problem due to the limits on the amount of time-integration we can perform with MD.

In order to reduce these sampling problems, the concept of a non-physical transition pathway between similar states is generally employed. This follows from realization that free energy is a state function, and therefore can be calculated by any pathway, even a non-physical one, as long as the endpoints correspond to the states in which you are interested. To this end, a new variable, λ, is introduced to the analytic potential function, subject to the boundary conditions $V(\lambda=0) = V_a$ and $V(\lambda=1) = V_b$. The exact form of the λ dependence, which is discussed elsewhere [19], is such that as λ increases from 0 to 1, the potential function monotonically gains more of the 'character' of the pure 'b-state' potential (and correspondingly loses more of the 'character' of the pure 'a-state' potential). The free energy difference between two physical states 'a' and 'b' is then calculated as the sum of free differences between these more closely spaced 'λ' states:

$$\Delta G = G_b - G_a = \sum_{i=1}^{N} \Delta G_i \tag{5}$$

where

$$\Delta G_i = -RT \ln \langle e^{-(V(\mathbf{r},\lambda_i)-V(\mathbf{r},\lambda_{i-1}))/RT} \rangle_{\lambda_{i-1}} \tag{6}$$

and $\lambda_0=0$ and $\lambda_N=1$. How to best choose the spacing $\delta\lambda_i$ between any two sequential λ states is a methodological question that has not been definitively answered. In most work a constant $\delta\lambda$ interval is used, though we have recently introduced another procedure that appears to be more efficient (dynamically modified windows [22]; see below).

The discussion above describes the FEP technique often termed 'window' growth. A second commonly-used technique, 'slow growth' or 'continuous integration', can be derived as a limiting case of window growth, where the $\delta\lambda$ spacings are so small that any two sequential λ states are extremely similar. We then assume that the ensemble average used to calculate any ΔG_i can be approximated by a single, instantaneous value:

$$\begin{aligned}\Delta G_i &= -RT \ln \langle e^{-(V(\mathbf{r},\lambda_i)-V(\mathbf{r},\lambda_{i-1}))/RT} \rangle_{\lambda_{i-1}} \approx -RT \ln e^{-(V(\mathbf{r},\lambda_i)-V(\mathbf{r},\lambda_{i-1}))/RT} \\ &= V(\mathbf{r},\lambda_i) - V(\mathbf{r},\lambda_{i-1})\end{aligned} \tag{7}$$

and so (from Eq. 5)

$$G_b - G_a = \Delta G = \sum_{i=1}^{N} \Delta V_i \tag{8}$$

where $\Delta V_i = V(\mathbf{r},\lambda_i) - V(\mathbf{r},\lambda_{i-1})$.

Another statistical mechanical derivation of slow growth can be used yielding the equation

$$G_b - G_a = \Delta G = \sum_{i=1}^{N} \frac{\partial V}{\partial \lambda} \partial \lambda_i \tag{9}$$

with the derivative evaluated for $\lambda = \lambda_i$. This expression is equivalent to Eq. 8, using the differential approximation (though the numerical precision of the two formulations may differ).

Below we discuss applications, observations, and pitfalls associated with the implementation of these equations. All calculations described were performed with an extensively modified version [22] of program AMBER/GIBBS 3.0 [23].

Free Energy Perturbation Calculations: Methodological Choices and Pitfalls

Checks on insufficient sampling and fortuitous results

One of the biggest advantages of FEP calculations - that they yield a single experimentally-relevant value - is also one of their biggest shortcomings. Because given enough tries, getting a 'good' value from a free energy calculation is not that hard, but is frequently fortuitous. Making small, reasonable changes in the simulation conditions will change both the MD trajectory and, if convergence has not been reached, the calculated free energy.

Additional testing is required to ensure that the calculated value is converged/ legitimate. Running a simulation for a longer period of time should yield a more accurate evaluation of the free energy, and simulations run for different lengths of time should yield approximately the same value if the shorter simulation has converged. If they do not, the shorter simulation, and perhaps the longer simulation as well, have failed to realize an accurate value. The lack of such confirmatory simulations is one of the most glaring shortcomings of many simulations in the literature.

Figure 1 shows an example of a case where a very short simulation gave a fortuitously excellent (compared to experiment) value for the free energy, a fact pointed out by the significantly worse value obtained by a slightly longer simulation run under the same conditions.

Machine precision and convergence

Since the MD master equation (Eq. 3) is integrated using an approximate method (e.g., 'leapfrog', Verlet, etc.), the calculated trajectory will be affected by the precision of the forces calculated at each intermediate time-point (energy evaluation). Even very small differences in the precision at each step can result, after several hundred or thousand steps, in much different trajectories for the same system with the same starting conformation.

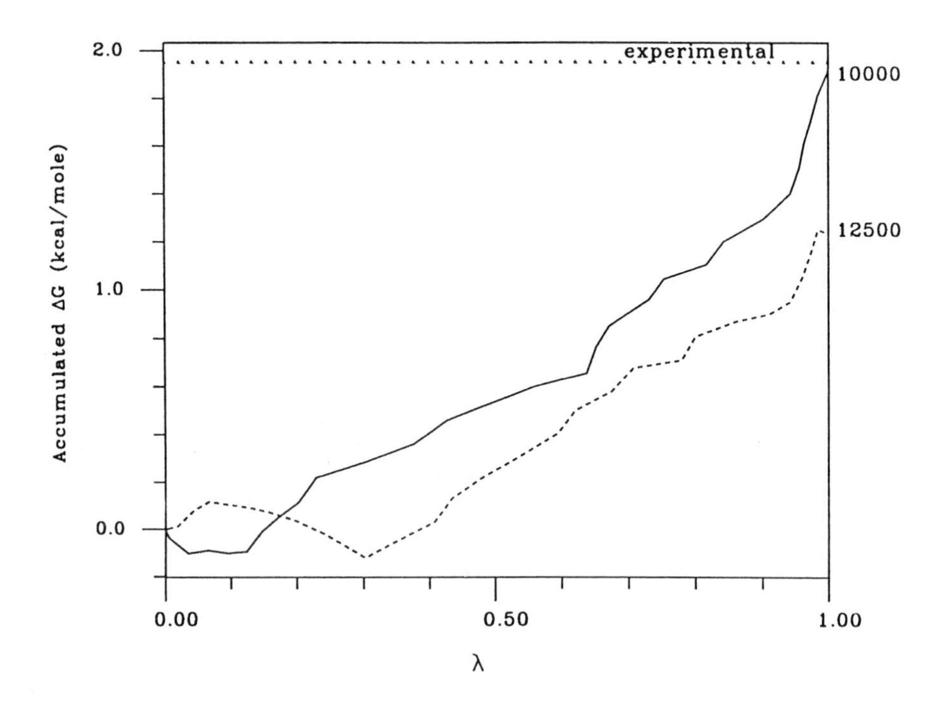

Fig. 1. Obtaining a seemingly 'good' value from a FEP calculation can be fortuitous. Additional testing is required to ensure the value is legitimate. Here the accumulated free energies are plotted versus λ for the simulations of the process nothing → methane in water. One, run for 10 000 steps (10 ps), seems to give excellent agreement with the experimentally observed value. But when the same simulation is run somewhat longer/ more accurately (12 500 steps/12.5 ps), a much worse result is obtained - implying the agreement with experiment for the shorter simulation was fortuitous. Both simulations were run with a periodic box of 295 water molecules, at constant pressure (~1 bar) and temperature (300 K), and with a non-bonded cutoff of 8 Å. Standard slow growth was used. The experimental value is from Ref. 27.

If the calculation has not converged, these differences in trajectory can translate into significantly different calculated free energies. Consider, for example, Fig. 2A. When writing FORTRAN programs to run on the Cray supercomputer [24], one can either calculate additive sums in-line (as normal), or one can call a routine SSUM which calculates these sums using a (nominally) faster vectorized set of instructions. The two methods supposedly give the same results out to the machine (64 bit) precision of the calculation. However, they apparently handle the least significant bits of precision somewhat differently, because free energy calculations performed using SSUM calls substituted for the in-line adds at only two places in the program (force accumulation loops) give divergent trajectories.

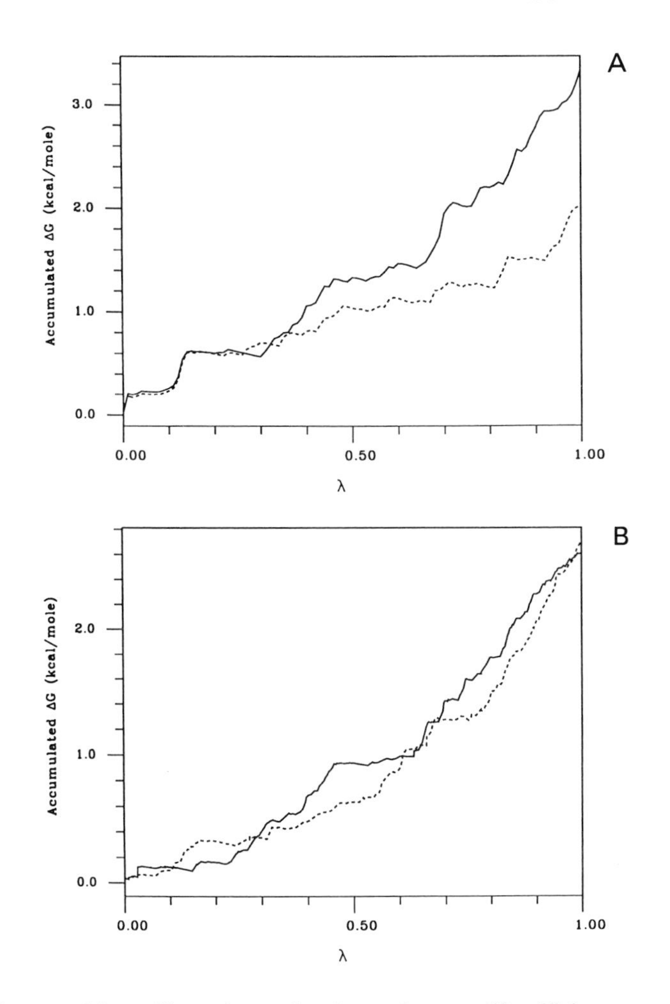

Fig. 2. Machine precision affects the molecular trajectory. If sufficient sampling is not carried out, and consequently convergence is not reached, the free energies calculated using differing trajectories can be very different. (A) Accumulated free energy versus λ plotted for the simulation nothing → neon in water. The entire simulation was run in 10 ps (10 000 steps), too little time to calculate a converged value. In this case, an extremely minor change in the precision of the program [in-line sums (solid line) to SSUM calls (dashed line); see text] results in widely divergent calculated free energies. (B) Accumulated free energy versus λ plotted for the simulation nothing → neon in water on both the Cray (solid line) and FPS (dashed line) computers. Although the precision difference between the Cray and FPS is much larger than the difference described for (A), nearly identical results are obtained when the simulation is run to near-convergence. All simulations were run with a periodic box of 295 water molecules, at constant pressure (~1 bar) and temperature (300 K), and with a non-bonded cutoff of 8 Å. For (A), normal slow growth was used; for (B), the dynamically modified windows method was used.

Figure 2A shows that if the calculation has not been run long enough for the convergence, the calculated free energies can be much different, *even for as small a change as this in the calculation.*

Figure 2B demonstrates the opposite effect. The same simulation with the same program code was run on both Cray and FPS [25] computers. While both calculations were run with 64-bit precision, it is well known that the actual numerical precision of the two computers differs somewhat, and significantly more than the precision difference between SSUMs and in-line sums. Nonetheless, when the same simulation as in Fig. 2A was run long enough for convergence to be obtained, the results on the two computers are nearly identical.

Small reasonable modifications of a system can affect the calculated free energy

There are many systematic corrections made in the course of an MD simulation. Corrections of coordinates, velocities, and/or the periodic box size are made to simulate a desired ensemble (NVT, NPT, etc.). Though reasonable equations for the smooth implementations of these changes have been derived [26], it remains to be conclusively shown that the changes do not affect FEP calculations, which are particularly sensitive to trajectory changes. As a gross example, consider Fig. 3. There, the effects of removing the center of mass motion of the krypton atom in a nothing → krypton simulation in a periodic box of water is shown. Removal of center-of-mass motion is frequently performed in normal MD calculations. But clearly, the trajectory of the atom has been affected by the motion removal, and this is reflected in the free energy curve. Additional equilibration is probably required after the center-of-mass removal. Though the effects of constant temperature, pressure, etc. algorithms on the calculated energies are expected to be much more subtle, it is worthwhile to note their potential influence on these values.

The non-bond mixing rules can affect the rate of convergence

Normally, the mixed optimum van der Waals contact distance between two particles i and j (see Eq. 1) is calculated as the arithmetic sum of the individual van der Waals radii R^*_{ii} and R^*_{jj}:

$$R^*_{ij}(\lambda) = R^*_{ii}(\lambda) + R^*_{jj}(\lambda) \tag{10}$$

For most cases, this is fine. But when one of the interacting particles 'disappears' at one of the physical endpoints [e.g., $R^*_{ii}(\lambda = 0) = 0$], then the normally-used λ dependencies yield

$$R^*_{ij}(\lambda = 0) = R^*_{ii}(\lambda = 0) + R^*_{jj}(\lambda = 0) = R^*_{jj}(\lambda = 0) \tag{11}$$

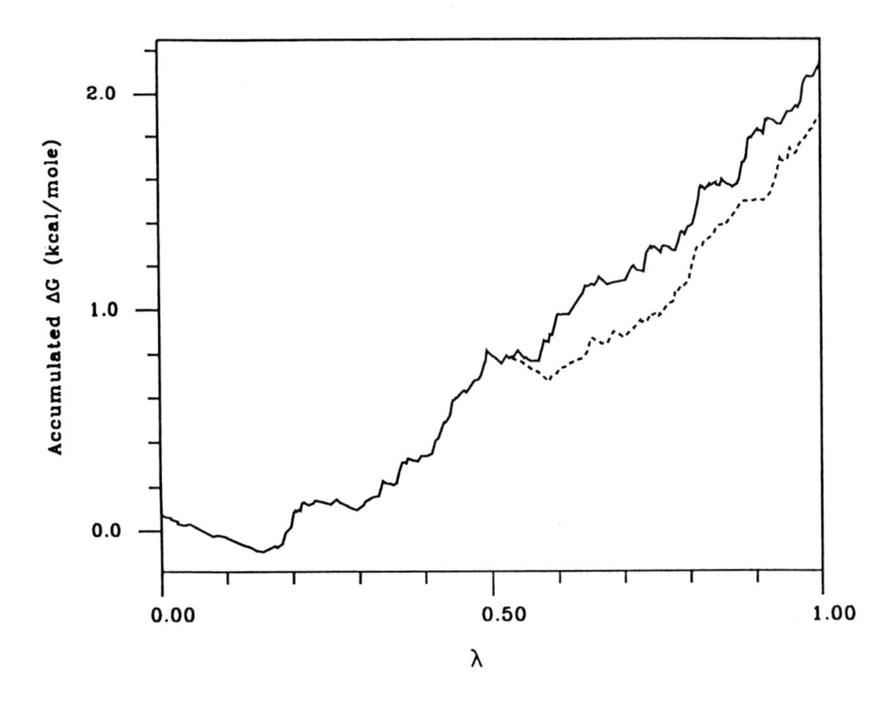

Fig. 3. Removal of the center-of-mass motion is frequently performed in normal MD calculations. But this affects the calculated trajectories and (if the simulation is not run slowly enough) free energies. Here the accumulated free energies for the simulation nothing → Kr in water are plotted versus λ for calculations run without removing center-of-mass motion (solid line), and removing this motion once at around $\lambda = 0.52$. Both simulations were run with a periodic box of 295 water molecules, at constant pressure (~1 bar) and temperature (300 K) conditions, and with a non-bonded cutoff of 8 Å. The dynamically modified windows technique was used.

If particle j is the same at both endpoints, then $R^*_{jj}(\lambda=0) = R^*_{jj}$. In other words, even at the endpoint where one of the particles has vanished, the equilibrium distance between the particles never becomes less than R^*_{jj}. Correctly sampling the potential energy difference for a change where a particle largely exists in one state but not in the other is particularly difficult for the reasons we discussed earlier. A better converging model might therefore be one where the equilibrium distance between the particles shrinks down to a small value as we near the physical state where one of particles has vanished.

Figure 4 plots the difference between the accumulated free energy curves for the simulation nothing → neon (in water) using both normal mixing (solid curve) and modified mixing, with $R^*_{ij}(\lambda=0) = 0.20$ Å. Elsewhere, we have obtained good results for nothing → methane using a value of $R^*_{ij}(\lambda=0) = 0$ [22].

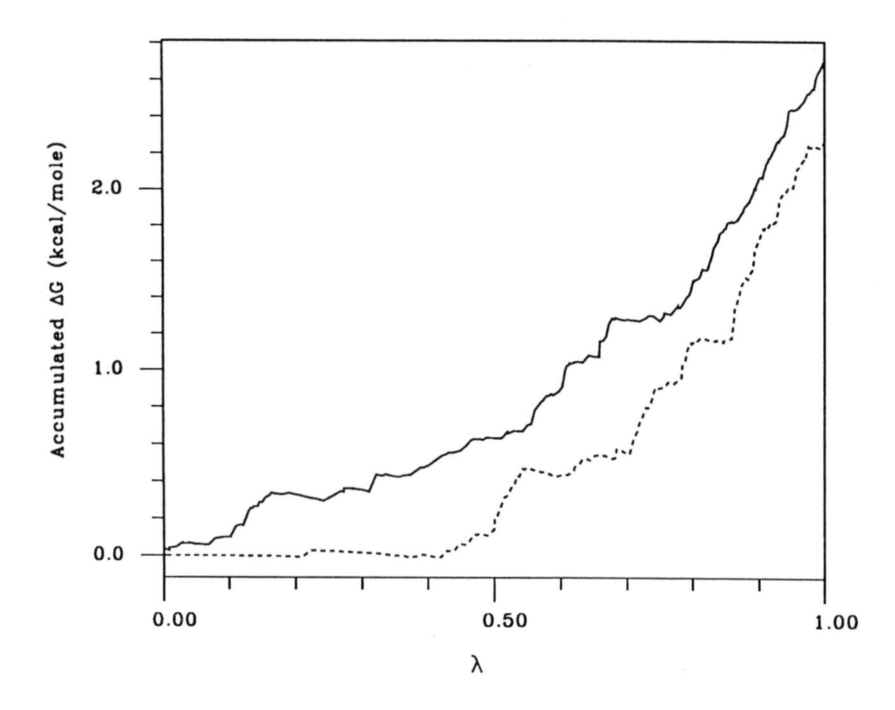

Fig. 4. The mixing rules generating R^*_{ij} for use in the non-bonded potential can affect the rate at which a free energy simulation can accurately be carried out. When one interacting particle vanishes, normal mixing rules give $R^*_{ij} = R_{jj}$ (see Eqs. 10 and 11). As described in the text, in many ways it makes more sense to set R_{ij} at this point to a small value. Here the effects of normal mixing (solid line) and modified mixing with $R^*_{ij} = 0.2$ Å (dashed line) are shown for the simulation nothing → neon in water. Both simulations were run with a periodic box of 295 water molecules, at constant pressure (~1 bar) and temperature (300 K), and with a non-bonded cutoff of 8 Å. The dynamically modified windows technique was used.

Effects of the integration time-step on calculated free energy

As we have noted, the complexity of the potential function precludes an exact integration of the MD master equation (Eq. 3). Instead, a predictor algorithm must be used. Several exist, but all share the feature that they predict the coordinates at time $t + \delta t$ from the forces at time t [21]. By increasing δt we can increase the rate at which we sample new conformations. But, unfortunately, the integration algorithms break down if δt is too large. For this reason, the maximum δt one can use is limited to the approximate range $0.5 \text{ fs} \leqslant \delta t \leqslant 2 \text{ fs}$.

For time-steps grossly too large, the effects are obvious: the dynamics trajectory becomes unstable, causing wild fluctuations in the temperature (velocities) and coordinates. However, it is unknown if certain time-steps that do not result in an obviously unstable trajectory may still disturb the integration in more subtle ways that may be reflected in the calculated free energies.

Table 1 gives the results of free energy calculations on three noble gases in water, using time-steps of both 1 fs and 2 fs. As can be seen, the results using the two time-steps differ, though they are within the estimated errors of one another. Interestingly, the results using a 1 fs time-step are closer to experiment for neon and krypton than those for the 2 fs runs (which should have sampled approximately twice as much time-space). While the results for xenon are closer to experiment for the 2 fs calculation, there is reason to believe that the parameters used in the calculation may not, in fact, lead to a converged, predicted value close to experiment (see Ref. 22). In all cases, the hysteresis for the 1 fs time-step run is better than, or about the same as, that for the corresponding 2 fs run.

Table 1 *Free energies of solvation for noble gases*[a]

Solute	Calculated	Experimental[b]	Steps[c]
1.0-fs time-steps			
Neon	2.51 ± 0.19	2.48	20 559
Krypton	1.65 ± 0.39	1.66	21 556
Xenon	1.05 ± 0.28	1.45	20 477
2.0-fs time-steps			
Neon	2.22 ± 0.39	2.48	19 077
Krypton	1.49 ± 0.37	1.66	19 157
Xenon	1.48 ± 0.39	1.45	20 278

[a] Values in kcal/mol. Calculations were performed in a periodic box of 295 water molecules at constant pressure (~1 bar) and temperature (300 K) conditions. The dynamically modified windows technique was used. Reported error ranges represent the hysteresis for simulations run forward and backward (λ $0 \rightarrow 1$ and $1 \rightarrow 0$).

[b] Experimental values taken from Ref. 30.

[c] Number of steps is averaged over the forward and reverse simulations.

At any rate, it is clear that calculated free energies show at least some dependence on the time-step used, and it remains to be conclusively determined whether simulations with time-steps at the high end of the usable range (i.e., 2 fs) produce simply better converged results, or results that harbor errors due to subtle problems with the integration formulae.

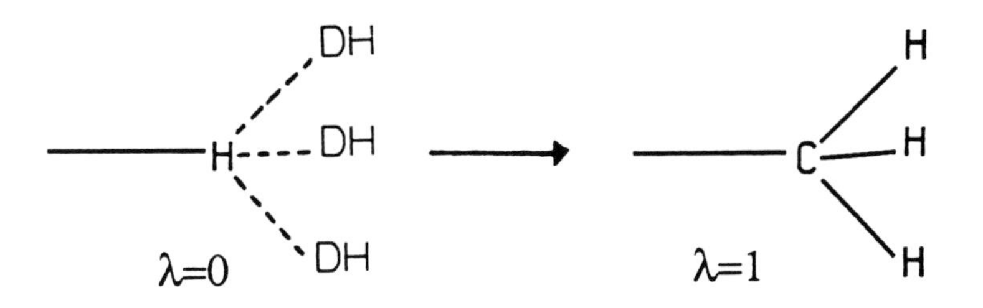

Fig. 5. Schematic description of the form a perturbation takes in AMBER. The perturbation shown changes a hydrogen substituent to a methyl group. The non-bonded parameters of the dummy hydrogens (DH) are zero, and are changed into those of real hydrogens (H) as λ goes from 0 to 1. The connectivity of the group does not change with λ. How the internal coordinates of the group change with λ, particularly the equilibrium bond lengths r_{eq}, is an important methodological question.

Formulation of $r_{eq}(\lambda)$ for bonds to 'vanishing' atoms

Another question of methodology that affects the rate of convergence of FEP calculations is that of how r_{eq}, the equilibrium bond distance (Eq. 1) changes with λ for a bond to an atom that 'vanishes' at one of the physical endpoint states. Figure 5 shows the form that such a change takes in program AMBER, in the case of a hydrogen atom subsituent ($\lambda = 0$) changing into a methyl group ($\lambda = 1$). The connectivity of the group does not change with λ; only the parameters associated with the internal coordinates and non-bonded terms change. The value of r_{eq} at $\lambda = 1$ is clearly just the standard value for a C-H bond. In many simulations, r_{eq} at $\lambda = 0$ is also set to $r_{eq,C\text{-}H}$, resulting in bond lengths that do not change as the 'H-DH' bond changes into a 'C-H' bond. But consequently, in the first $\delta\lambda$ step away from $\lambda = 0$, the methyl group goes from being completely invisible to being visible, with large bond lengths. This means the partial hydrogen atoms may immediately protrude far into the surrounding solvent structure (though at the first step their non-bonded parameters will still be rather 'soft').

A second possible methodology, although one rarely used in the literature, assigns a small value to $r_{eq,H\text{-}DH}$ at $\lambda = 0$. The value of r_{eq} is then linearly increased so that it attains the value $r_{eq,C\text{-}H}$ at $\lambda = 1$. Since the attached hydrogens are closer to the central atom in the initial stages (low λ values) of the simulation, the surrounding solvent structure is disturbed more slowly. The free energy may consequently increase more slowly in the early stages of appearance of the methyl group, aiding convergence. Of course, these methodological choices must be made for any atom which 'disappears' in one of the physical states, not just a methyl group.

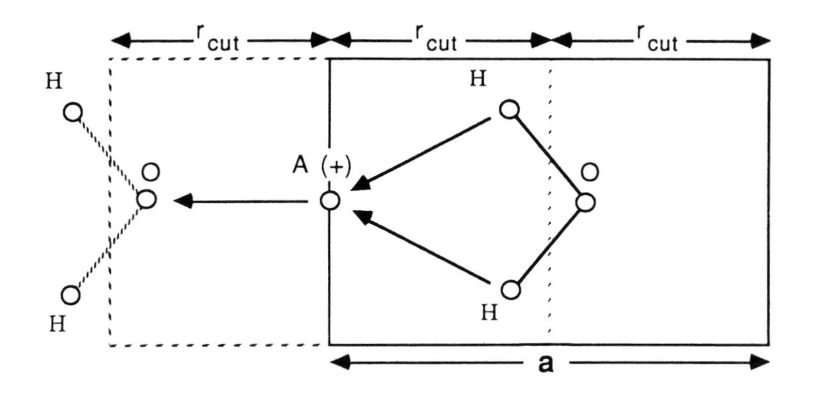

*Fig. 6. Schematic view of the problem with using too small a value for the non-bonded cutoff in AMBER. The periodic box is shown (in two dimensions) by the solid lines. Half of one of the periodic images of the box is shown by the dashed lines to the left. A positively charged atom (A) is shown at the left side of periodic box. A water molecule is shown towards the center of the box, and one of its periodic images is shown to the left. If the cutoff, r_{cut} is set to half of the length of the box, **a**/2, then atom (A) interacts with the hydrogen atoms of the water molecule, and with the image of the oxygen atom of this molecule. The result is three forces on atom (A) all in the same direction, rather than the correct near-cancellation of forces that would be calculated if interactions only with the parent water molecule (or only with its image) were used. This arises from the fact that the non-bonded cutoff is applied on an atom-atom basis (not a residue-residue basis) in the energy calculation routines of AMBER, and that a minimum image convention is used for periodic boundary conditions. The problem can be avoided by setting $r_{cut} \geq \mathbf{a}/2 + r_{solvent}$, where $r_{solvent}$ is the width of the solvent molecule (in this case water). For programs where the non-bonded cutoff is applied on a residue-residue basis, a cutoff of $r_{cut} \geq \mathbf{a}/2$ is sufficient.*

We have calculated [22] the free energy for the change nothing → methane in water, both with $r_{eq,H\text{-}DH} = r_{eq,C\text{-}H} = 1.09$ Å, and with $r_{eq,H\text{-}DH} = 0.2$ Å. The simulations were run for equivalent numbers of steps using the dynamically modified windows technique. We calculate free energies of 2.44 ± 0.44 and 2.09 ± 0.09 kcal/mol, respectively, for the two protocols. The experimental value is 1.93 kcal/mol [27]. Thus in this case, at least, our results imply that allowing bonds to 'dummy' atoms to shrink seems to lead to better convergence in the same number of steps. Additional testing is required to reach a definitive conclusion.

It should be noted that while it is possible in AMBER to choose a separate value of r_{eq} for the group which is appearing from nothing, not all implementations of FEP calculations allow the user to do this. For example, in some implementations both the original group and the group to which it is being perturbed exist for all values of λ, and the internal coordinate parameters do not change with λ. The two groups do not 'see' each other, and the non-bonded parameters for each are scaled so that as one group 'disappears' from the system, the other group 'appears'. Given our tentative results favoring the shrinkage of bonds, it is worth evaluating whether this second type of implementation can be improved.

Cutoffs...and caveats

The most computer-intensive part of the energy calculation for a large system is generally the non-bonded energy. The number of possible non-bonded pairs increases as $N(N-1)$, where N is the number of atoms in the system. For this reason, it is frequently desirable to use a cut-off distance. Atom pairs separated by more than the cut-off distance are considered to contribute negligibly to the calculated energy, and so their interactions are ignored.

This is clearly an approximation. Certainly, point charge electrostatic interactions (Eq. 1), which decrease with the inverse of the interatomic distance, will only be negligible beyond a very large cutoff. For example, for a cutoff of 8 Å (a commonly used value), two charges of 0.3 e will still have an interaction energy of 3.7 kcal/mol. Even though the gross neutrality of groups will generally lead to near-cancellation of electrostatic effects for distances larger than the cutoff, it is clear that the choice of cutoff introduces some amount of inaccuracy to calculations. Whether the inaccuracy is acceptably small for FEP simulations has yet to be satisfactorily tested.

Another possible pitfall associated with cutoffs can occur in programs such as AMBER, when using periodic boundary conditions. In these programs, non-bonded interactions are evaluated on an atom-atom 'minimum image' basis. As shown in Fig. 6, if the cutoff (r_{cut}) is such that

$$r_{cut} \geq (\mathbf{a}/2 - r_{solvent}) \tag{12}$$

where $\mathbf{a}$ is the box length and $r_{solvent}$ is the width of a solvent molecule, it is possible that interaction between a solute atom and a solvent molecule can be split so that a large unrealistic force is calculated, resulting in an incorrectly evaluated trajectory and pressure (because the virial, $-\frac{1}{2}\Sigma \mathbf{r} \cdot \mathbf{F}$, will be wrong). Consequently, the density of the solution and the calculated free energy will be erroneous (Table 2). This can be avoided by always ensuring that $r_{cut} \leq$ (minimum box dimension)/2−(width of solvent molecule). In some programs, the non-bonded interactions are evaluated on a *residue-based* 'minimum image' basis. In these cases, it is only necessary that $r_{cut} \leq$ (minimum box dimension)/2.

In any case, the dimensions of the periodic box should be checked after start of the simulation to ensure the restrictions on r_{cut} are still fulfilled.

Table 2 *Effects of improper cutoffs in periodic simulations*[a]

Cutoff (Å)	Free energy (kcal/mol)	Volume (Å^3)	Density (g/cm^3)
8	3.19	8899	1.009
9	3.01	9124	1.038
20	5.91	7120	0.807

[a] Values reported for the simulation nothing → neon run in a periodic box of 295 water molecules at constant pressure (~ 1 bar) and temperature (300 K) using normal slow growth. Volumes are averaged over the simulations. Assuming a pure water density of 1 g/cm^3, a symmetric box of 295 water molecules should be approximately 20.7 Å in each dimension, and have a volume of 8820 Å^3. Using Eq. 12, with $r_{water} \approx 1.5$ Å, give a maximum safe cutoff of 8.85 Å. The latter two simulations listed used cutoffs exceeding this, and the resulting densities and free energies are unreliable.

How should λ be changed?

In the past, the rate of change of λ has generally been fixed over large ranges of λ (often over the entire range 0 → 1). It seems more reasonable to change λ so that when the free energy is changing quickly with λ, λ is changed slowly, and vice versa. In a recent paper [22], we described the implementation and use of a method, Dynamically Modified Windows, which does exactly this. In the method, the local slope of the free energy versus λ curve is used to determine the next value of δλ.

The energies of solvation for methane and several noble gases were calculated using the new method and standard (fixed δλ) slow growth for the same numbers of steps. A comparison of the results presented in the paper indicates quite strongly that the new method allows converged free energies to be calculated in many fewer steps than standard methods.

Other questions

We have discussed and presented preliminary data for a number of issues relevant to free energy perturbation calculations. Still many other questions remain unanswered, among them:

(1) How good is the potential energy force field for FEP calculations? Is the analytic form acceptably accurate? How about the parameterization? Are current solvent models adequate?

(2) Is MD the optimal method for evaluating the ensemble average? Monte Carlo methods can also be, and have been, used to generate data applicable to such purposes [28].

(3) How do various simulation conditions affect calculated free energies and the rate of convergence: (a) Periodic boundary conditions (constant pressure and volume algorithms); (b) SHAKE [29] (a commonly-used algorithm for

constraining bond lengths, thereby allowing somewhat bigger time-steps to be used)?

(4) How do the various formulations of FEP simulations compare: 'windows' versus 'thermodynamic integration' versus 'slow growth'?

(5) How can we derive reasonable estimates of the errors in our results?

Conclusions

Unquestionably, FEP simulations hold great promise for allowing us to finally predict many important molecular properties. But before the potential of the method can be fully realized, we must derive an *effective, consistent and reliable* methodology for carrying out these calculations. As we have illustrated in this paper, there are a number of issues which still need to be resolved regarding FEP simulations. Until this has been done, results from calculations using FEP methods should be examined with a healthy amount of scrutiny.

Acknowledgements

We would like to thank the NSF (DMB-87-14775) and the San Diego Supercomputing Center for supercomputing time. We would also like to thank the NIH, The National Cancer Institute (CA-25644), DARPA (grant N00014-86-K-0757, R. Langridge P.I.) and Merck, Sharp and Dohme for research support.

References

1. Mezei, M. and Beveridge, D.L., Ann. N.Y. Acad. Sci., 482(1986) 1.
2. Bash, P.A., Singh, U.C., Langridge, R. and Kollman, P.A., Science, 236(1987) 564.
3. Schaefer III, H.F. (Ed.) Modern Theoretical Chemistry, Vol. 3 (Methods of Electronic Structure Theory), Plenum, New York, 1977.
4. Schaefer III, H.F. (Ed.) Modern Theoretical Chemistry, Vol. 4 (Methods of Electronic Structure Theory), Plenum, New York, 1977.
5. Segal, G. (Ed.) Modern Theoretical Chemistry, Vols. 7-8 (Semiempirical Methods of Electronic Structure Calculation), Plenum, New York, 1977.
6. Ramachandran, G.N. Ramakrishnan, C. and Sasisekharan, V., J. Mol. Biol., 7(1963) 95.
7. Lifson, S. and Warshel, A., J. Chem. Phys, 49(1968) 5116.
8. Allinger, N., J. Am. Chem. Soc., 107(1977) 4079.
9. Weiner, S.J., Kollman, P.A., Case, D.A., Singh, U.C., Ghio, C., Alagona, G., Profeta, S. and Weiner, P., J. Am. Chem. Soc., 106(1984) 765.
10. Weiner, S.J., Kollman, P.A., Nguyen, D.T. and Case, D.A., J. Comput. Chem., 7(1986) 230.
11. Pettitt, B.M. and Karplus, M., In Burgen, A.S.V., Roberts, G.C.K., Tute, M.S., (Eds.) Molecular Graphics and Drug Design, Elsevier, New York, 1986.

12. Van Gunsteren, W.F. and Berendsen, H.J.C., The power of dynamic modelling of molecular systems, to be submitted to J. Comput.-Aided Mol. Design.
13. Howard, A.E. and Kollman, P.A., J. Med. Chem., 31 (1988) 1669.
14. McCammon, J.A., Gelin, B. and Karplus, M., Nature, 267 (1976) 585.
15. Postma, J.P.M., Berendsen, H.J.C. and Haak, J.R., Faraday Symp. Chem. Soc., 17 (1982) 55.
16. Warshel, A., J. Phys. Chem., 86 (1982) 2218.
17. Jorgensen, W.L. and Ravimohan, C., J. Chem. Phys., 83 (1985) 3050.
18. Lybrand, T., McCammon, J.A. and Wipff, G., Proc. Natl. Acad. Sci., U.S.A. 83 (1986) 833.
19 Singh, U.C., Brown, F.K., Bash, P.A. and Kollman, P.A., J. Am. Chem. Soc., 109 (1987) 1607.
20. Zwanzig, R.W., J. Chem. Phys., 22 (1954) 1420.
21. Karplus, M. and McCammon, J.A., Annu. Rev. Biochem., 52 (1983) 263.
22. Pearlman, D.A. and Kollman, P.A., J. Chem. Phys., in press.
23. Singh, U.C., Weiner, P.K., Caldwell, J.W. and Kollman, P.A., AMBER (UCSF), version 3.0 Department of Pharmaceutical Chemistry, University of California, San Francisco, CA, 1986.
24. Cray XMP. Cray Research, Mendota Heights, MN.
25. FPS-264. Floating Point Systems, Portland, OR.
26. Berendsen, H.J.C., Postma, J.P.M., van Gunsteren, W.F., Nola, A.D. and Haak, J.R., J. Chem. Phys., 81 (1984) 3684.
27. Ben-Naim, A. and Marcus, Y., J. Chem. Phys., 81 (1984) 2016.
28. Jorgensen, W.L., J. Phys. Chem., 87 (1983) 5304.
29. Van Gunsteren, W.F. and Berendsen, H.J.C., Mol. Phys., 34 (1977) 1311.
30. Franks, F. and Reid, D.S., In Franks, F. (Ed.) Water - A Comprehensive Treatise, Vol. 2, Plenum, New York, 1973.

Microscopic free energy calculations in solvated macromolecules as a primary structure-function correlator and the MOLARIS program

Arieh Warshel and Steve Creighton
Department of Chemistry, University of Southern California, Los Angeles, CA 90089, U.S.A.

1. Introduction

The field of computer simulation of biological molecules has progressed enormously in recent years. This progress has been strongly driven by the increased availability of affordable computer time. Despite this rapid progress it seems that the emphasis on biologically relevant problems and on the issue of structure-function correlation has progressed quite slowly. In fact, until the recent emergence of genetic engineering, it seems that only few studies have attempted to evaluate the free energies of actual biological processes using microscopic models [1]. Nevertheless, we are now witnessing an increased understanding in the scientific community of the fact that free energies (and in particular the electrostatic self-energy or 'solvation energy') are the key correlators between structure and function.

Realizing quite early the crucial importance of free energy calculations, we faced major practical problems. Experience with the enormous dimensionality of even medium-sized molecules [2] made it clear that the computers of the 70s could not be used to sample the phase space of proteins in any meaningful way (regardless of the method used) and that the only feasible calculations were energy minimizations that could be used to explore local minima and to relax steric strain [3]. Yet, repeated computer experiments with large molecules indicated that steric forces are almost always relaxed and that the key factor that determines free energy of reacting substrates or cofactors is their electrostatic interaction with the surrounding enzyme and water system. It also became clear that this energy with its long range features must be modeled by taking into account *all* its key components (which could be accomplished only by simplified approximations) rather than dealing rigorously with few components (e.g., treating an ion and few water molecules quantum mechanically). The resulting model, which will be reviewed in Section 2, appeared to be the method of choice for free energy calculations in the late 70s. The progress in computer power in the 80s led us and others to the implementation of more rigorous and much more expensive approaches that will be considered in Section 3.

In this paper we will attempt to describe our perspectives and early contributions to free energy calculations in proteins arguing that rigorous models might be quite meaningless unless they could be implemented in a converging procedure with the available computer technology. Including only part of the key factors in rigorous models is quite ineffective in analyzing a given system. On the other hand focusing on relevant factors allows one to progress at the same rate as the progress in hardware technology. This will be demonstrated by considering different models for free energy calculations and their performance. We will use this opportunity to consider our program package MOLARIS and its three main modules POLARIS, ENZYMIX and QCFF/SOL.

2. Simplified Microscopic Calculations of Free Energy in Solvated Proteins

True computer modeling of biological molecules should allow one to explore biological functions and specificity. Relevant problems about specificity can be formulated by free energy diagrams of the form presented in Fig. 1.

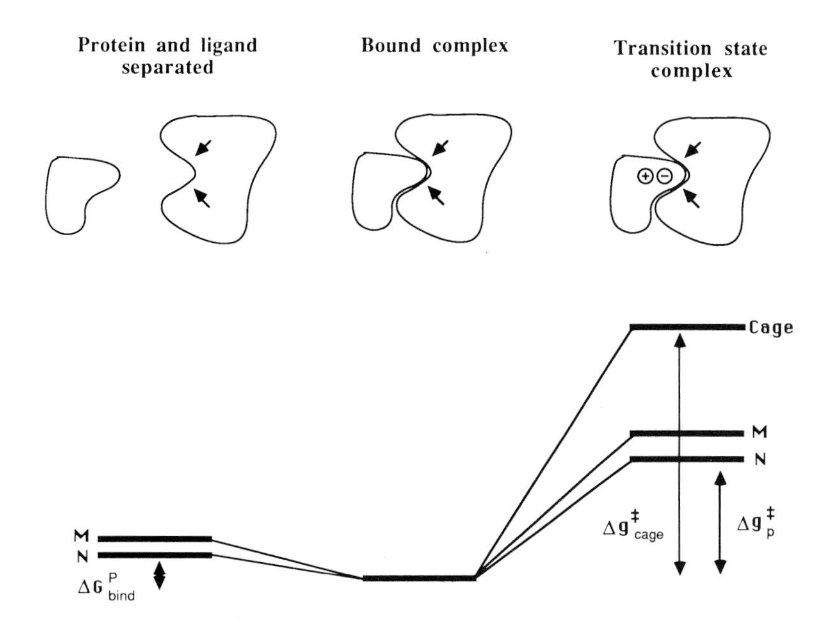

Fig. 1. Free energy contributions that determine protein specificity. N, M and cage designate, respectively, the reaction in the native protein, the mutant protein and in a reference solvent cage.

The ability to evaluate the 'solvation' free energy of the reacting system in proteins and in water appears to be the key for obtaining such energy diagrams and the key for quantitative structure-function correlation [1]. The problem is, however, to develop models capable of evaluating solvation free energies in proteins.

Realizing the above point quite early it became clear that all atom models would be impractical without further increase in computer power. Yet the electrostatic origin of solvation effects could be captured by simpler models. The obvious candidates were the available macroscopic models [4-6]. For example the energy of an ion pair with dipole moment μ in a protein can be described in (kcal/mol) by

$$\Delta G = - \frac{166\mu^2}{b^3} \frac{2\epsilon - 2}{2\epsilon + 1} \tag{1}$$

where b is the radius of the protein cavity around the ions and ϵ is the dielectric constant of the protein. Unfortunately, both b and ϵ are not defined by any macroscopic concept as the microscopic nature of the protein environment is not the same in different sites. The need for a microscopic model which is, however, compatible with the available computer power, led to the development of the protein dipoles-Langevin dipoles (PDLD) model [7, 8]. This model represents all the key electrostatic contributions of the protein (permanent dipoles and induced dipoles) and represents the surrounding water by a grid of Langevin-type dipoles (see Ref. 8 for details).

The philosophy behind this simplified water model is as follows. The average polarization of any given water molecules near an ion is related in some way to the field from the ion. If we knew the distribution function for this average polarization then we could evaluate the free energy of solvation of the ion by calculating the sum of the product of the ion electric field and the average polarization of the water along that field for each water molecule and add the free energy of polarizing the waters. Fortunately, one can determine the polarization of water molecules as a function of the applied field from microscopic simulations [8a] (and/or by refining the model by fitting calculated and observed solvation energies) and represent the induced dipoles by a Langevin-type expression. Such an approach, which is discussed in detail elsewhere [8], accurately models the main physics associated with the solvation free energies without the need for an explicit water model. As much as the protein is concerned, the PDLD approach can be considered as an attempt to evaluate the electrostatic free energy at the average protein structure (as given by the X-ray coordinates) rather than attempting to evaluate the Boltzmann average of the electrostatic free energy over all of the possible protein configurations.

The seemingly oversimplified PDLD model appears to be surprisingly effective in evaluating free energy cycles, giving semiquantitative results. This includes early nonphysical microscopic free energy cycles (Fig. 2) for energies of charges in proteins [9a], calculations of redox energy in cytochromes [9b] and, most importantly, in calculating the free energy of enzymatic reactions [1, 8]. This model demonstrated that the focus on solvation free energy may hold the key for studies of structure-function correlation. It also became clear from inspection of the different contributions to the calculated solvation energies that simplified models which take all effects into account are far superior to rigorous treatment of parts of the system, e.g., neglecting the water around the protein while treating the substrate by rigorous quantum mechanical approaches.

The PDLD method also appears to provide an extremely effective model for studies of solvation energies of medium-sized molecules in solutions (e.g., different amino acids in water). The PDLD model has evolved in recent years into a powerful tool for dealing with solvation of large macromolecules including large proteins, ion channels, and viruses. The program package which implements PDLD is called POLARIS. Solvation free energy calculations using POLARIS can be done in minutes on Alliant superminicomputers and our research shows that

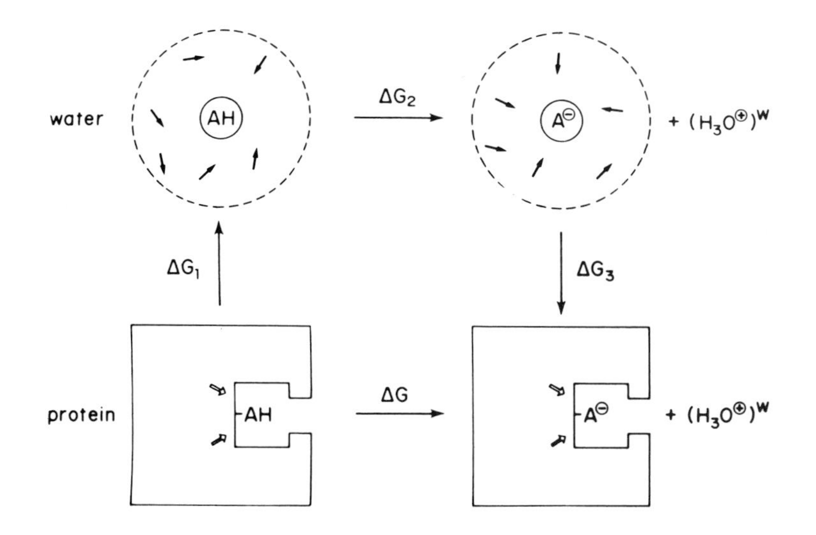

Fig. 2. Describing the thermodynamic cycle used to estimate the energetics of dissociation of an acidic group of a protein. The ΔG_i are given by $\Delta G_1 = G^{w}_{sol}(AH) - G^{p}_{sol}(AH)$, $\Delta G_2 = 2.3RT(pK^{w}_{a} - pH)$, and $\Delta G_3 = G^{p}_{sol}(A^{-}) - G^{w}_{sol}(A^{-})$. This figure is taken from Ref. 9a.

one can model the mean physics of solvation effects without the need of supercomputers.

2.1. The POLARIS program

2.1.1. The main features of POLARIS

POLARIS is a general program for performing quickly converging calculations of solvation free energies and electrostatic energies of molecules and macromolecules. The program is based on the PDLD approach [7, 8]. This approach represents explicitly all of the key electrostatic contributions in macromolecules (e.g., permanent dipoles, induced dipoles, and surrounding water molecules) but uses a simplified set of Langevin dipoles to represent the solvent - thus achieving an extremely fast convergence in the calculation of the solvation free energy.

The calculations involve a menu-driven preparation procedure using the program PREPARE where the system [composed of the macromolecule (or molecule) and water] is divided into three sections (see Fig. 3).

After preparing the regions, the user can define the thermodynamic cycle that he wishes to explore, (evaluating the pKa of an acidic group in a protein) and define the charges of the relevant reference groups in the two limiting states of Fig. 2.

Once the charges of the reference groups and the surrounding regions are defined, the actual PDLD calculations are performed by the program POLARIS. The program then provides an estimate of the following factors:

(i) $V_{Q\mu}$. This is the total electrostatic interaction between the residual charges (protein dipoles) in region II and the charges in region I.

(ii) $V_{Q\alpha}$. This is the energy of interaction between the induced dipoles in region II and the charges in regions I and II. The induced dipoles are determined by the self-consistent procedure described in Ref. 8b.

(iii) V_{QW}. This is the energy of interaction between the Langevin dipoles and the charges in regions I and II. This includes the free energy cost of polarizing the dipoles from their initial state of zero average polarization.

(iv) ΔG_{bulk}. This is an estimate of the contribution to the electrostatic free energy of solvation of region I from the bulk waters beyond the waters modeled by the Langevin dipoles.

(v) V_{QQ}. This is the sum of the electrostatic energy of each of the region I charges interacting with the other charges in region I.

The sum of all of the above interactions provides an estimate of the electrostatic free energy of the given system.

2.1.2. Specialized POLARIS tasks

POLARIS can be used (in addition to the standard PDLD mode described above) for additional related tasks. Some of these tasks are outlined below.

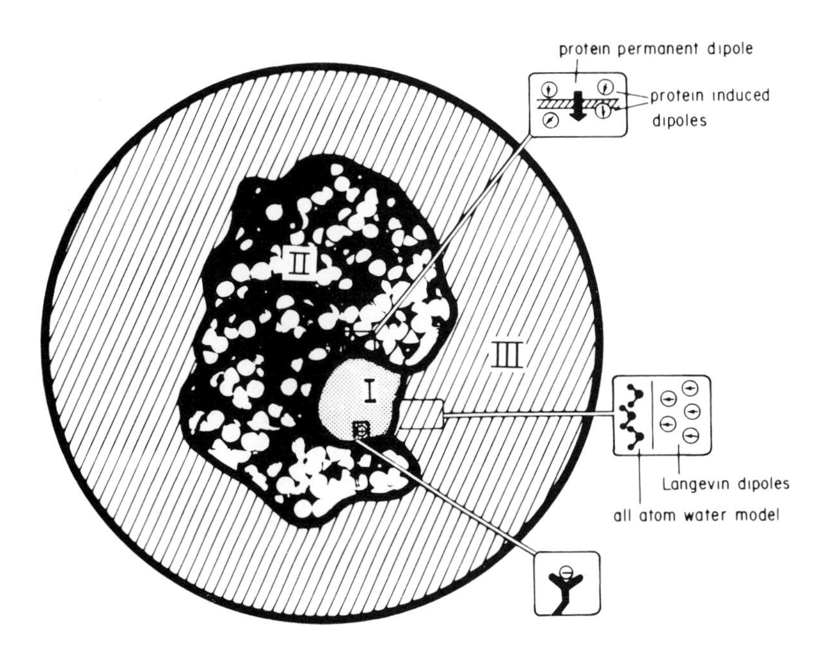

Fig. 3. The three-region model used to calculate the electrostatic energies in proteins. Region I includes the charged groups of interest, region II includes the permanent and induced dipoles of the protein and region III includes the surrounding water molecules, which can be simulated by an all-atom model for a few bound water molecules and by a Langevin-dipole model for the rest.

(i) Calculating electric fields. POLARIS can generate the coordinates of all of the protein atoms and Langevin dipoles in the system and the electric field vectors at each point. This field is due to the region I charges, the region II charges and the Langevin dipoles. This field can be either plotted or used for other purposes.

(ii) Calculating solvation energies of small molecules in water. It is possible to use the POLARIS program to calculate the solvation free energy of small molecules that are not proteins. The corresponding estimates, which are surprisingly reliable, are obtained almost instantly.

(iii) Incorporating PDLD in energy minimization studies. Frequently, one would like to augment the results of energy minimization studies or other conformation search procedures with solvation results. The simplest option is to add the PDLD solvation energy to the conformation energy of the system in question provided

by other energy minimization programs (or, much better, by MOLARIS or ENZYMIX).

(iv) Simulating discretized continuum models. The program can provide one with the useful results of macroscopic dielectric approaches without the need to solve the Poisson equation numerically (which customarily involves a large amount of computer time [11]).

The macroscopic results can be obtained by the user by performing a POLARIS calculation using a small grid size - typically 1 Å.

(v) Calculating protein electrostatic interaction with region I. By using the PREPARE program it is possible to calculate the electrostatic interaction of each protein residue in region II with the region I atoms. The program lists and plots the electrostatic contributions of different amino acids in a way that allows one to predict the effect of genetic modifications.

3. Calculations of Free Energies Using All-Atom Solvent Models

The advance in computer power in the past few years has made it possible to start sampling the phase space of multidimensional systems by brute force approaches such as Monte Carlo (MC) and molecular dynamics (MD) simulations. Formal statistical mechanical approaches which could not converge (and therefore were basically useless) in the 70s are now becoming the methods of choice in the 80s.

Our first attempt [10] to evaluate solvation free energies by an all-atom solvent model and MD simulations explored the effectiveness of the free energy perturbation (FEP) method [12, 13] which will be outlined below. In order to reduce the dimensionality of the problem we used the idea of surface constraints on the water drop (first introduced in the SCSSD model [14]) and used a limited number of water molecules surrounded by a surface of molecules that represents the effect of the surrounding in the corresponding infinite system (for our more refined SCAAS model and related models see Refs. 8, 15-17). Despite this saving in computer time, we encountered significant convergence problems (5 kcal/mol) for a job that took 10 h on the IBM 370. Smaller absolute errors were reported for nonpolar solutes [18]. Nevertheless, it became clear that such explicit calculation of free energies would require more computer power than was available in the early 80s.

Even with the limited computer power of that time, it was felt that non-converging FEP calculations could be used to verify the linear free energy relationships assumed in our PDLD calculations of enzymatic reactions [8]. This led to the implementation of FEP calculations in the evaluation of free energy changes between different intermediates of enzymatic reactions and the calculations of the corresponding activation free energies [19].

The real 'push' for free energy calculations in proteins did not occur until the emergence of genetic engineering of proteins: at site-specific mutagenesis experiments provided well-defined problems which are sufficiently simple to converge in favorable cases [20-24]. In order to consider our studies in this exciting field we will outline the main features of our specialized treatment of enzymatic reactions and related processes.

3.1. The ENZYMIX *program*

3.1.1. The purpose and philosophy of ENZYMIX

The main purpose of ENZYMIX is to generate the free energy profile of reactions in solution and in proteins. Comparison of the reaction profile for the solution reaction to that of the enzyme reaction allows the user to postulate various mechanisms for the enzymatic reaction. Using the EVB method it is possible to translate a postulated mechanism (e.g., proton transfer, nucleophilic attack, electron transfer, electrophilic attack, etc.) into a force field that the computer can understand and that can be used for calculation of the free energy profile of the postulated mechanism. Thus the computer allows the user to *simulate* the feasibility of his proposed mechanism by comparing the free energy profile of a given mechanism to the profiles of other mechanisms. The use of the ENZYMIX programs in *parallel* with experimental studies is a powerful way of determining the reaction mechanism since the programs are a source of ideas and the experiments are a check on the validity of the calculations. We feel that this is the most effective way to use ENZYMIX.

The philosophy behind ENZYMIX is that since it is *extremely* hard to generate accurate potential energy surfaces for even small molecules using the state-of-the-art ab initio techniques the study of enzymatic reactions must be done using simplified valence bond (VB) force fields - that accurately (but not exactly) reproduce the potential energy surface of the relevant reference reaction in solution and can be used to explore the *change* of the surface in the enzyme active site. These force fields are constructed using insights gained from both experimental and theoretical studies and they are *not at all* inferior to ab initio force fields, they merely represent an 'engineering' approach to molecular potential surfaces as opposed to the 'first principles' approach emphasized in ab initio methods. In particular, we insist on getting the most reliable *experimental* estimate of the gas phase energies of the charged fragments. When gas phase experiments are not available, then we use the experimental energies of forming the fragments in solution and the calculated solvation free energies of the fragments in solution, noting that (whether one likes it or not) it would be essential to recalibrate the current ab initio results on the same experimental free energies*. Since the force fields in ENZYMIX are *empirical* in nature it is possible to make them more accurate as more experimental information is gained.

3.1.2. What is the EVB method?

The EVB (empirical valence bond) method is a simple and effective way of including quantum mechanics into a FEP/MD simulation. This is very important since the modeling of chemical reactions requires a quantum mechanical treatment. The ability of bonds to move around during a reaction implies that there are more degrees of freedom in the chemical reaction than just the positions of the nucleii.

The program ENZYMIX represents the potential energy surfaces of proteins by a combination of a classical empirical force field and a quantum empirical valence bond force field. The classical force field is used to simulate the parts of the protein removed from the actual chemical reaction being studied since there is no bond breaking or making in this region. In the small region of the protein where there is a chemical reaction taking place, a quantum mechanical empirical method is used to represent the changing *electronic* (as opposed to nuclear) coordinates of the atoms involved in the reaction. A valence bond formalism is used to simulate the reacting atoms since this method is well suited to model bond making and breaking (the molecular orbital method is better suited to calculation of spectroscopic properties of molecules) and the empirical nature of the force field allows calibration of the model to accurately reproduce experiment.

The ground state potential energy surface (PES) is constructed by mixing (in a quantum mechanical sense) the properties of the different valence bond (VB) resonance structures that describe the chemical reaction that is taking place. Typically, the user will define a reactant and a product bonding pattern for the reaction that he is interested in and the EVB method allows the program to determine the energies and forces acting on the atoms that the user defines as quantum atoms as a function of not just the coordinates of the atoms in space but the percent character of reactant and product wavefunction in the actual wavefunction of the system. This dependence of the force field of the quantum atoms on the reactant and product character of the system alllows the user to cause the reaction to occur by slowly forcing the system to move from 100% reactant to 100% product wavefunction. As the quantum atoms are forced to react, the protein environment will attempt to 'follow' the reaction and it is possible to use the free energy perturbation formalism to determine

*This is by no means an attack on ab initio methods. Such methods provide a great deal of insight into chemical bonding but are at present useless for modeling truly huge systems like proteins.

the change in the *overall* free energy of the protein + reacting atoms + water system that the user is interested in [8, 12, 22].

3.1.3. Example: OH^- attack on a peptide C=O group

The attack of a hydroxyl molecule on the carbonyl carbon of a peptide group in the substrate of trypsin is readily simulated using these programs. In this example the system being studied and the relevant states are:

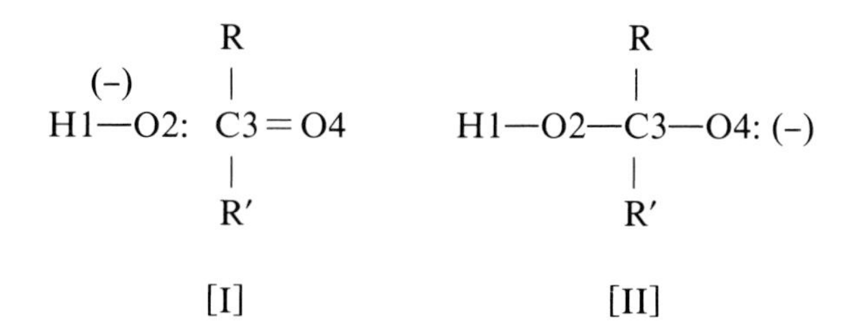

To simulate this process the user must first define the quantum atoms in the problem. Quantum atoms are any atoms that undergo a bonding change as the reaction progresses. In this example the quantum atoms are the two oxygen atoms (O2 and O4) and the carbon atom (C3) that is attacked. When the user defines the quantum atoms he must also define the type of the atoms in each resonance form and their charge in each resonance form. After a little thought and consultation of the ENZYMIX manual we decide:

Resonance form I		
Atom	Type	Charge
O2	Oh	-1.0
C3	Co	+0.3
O4	Oh	-0.3

Resonance form II		
Atom	Type	Charge
O2	Oh	-0.2
C3	Co	+0.2
O4	Oh	-1.0

Next the user must define the relevant structures in terms of the bonding patterns in each of the resonance forms. In the above examples there are two resonance forms. In resonance form I there is a bond between atom C3 and O2 and in resonance form II there are bonds between atoms O2 and C3 and between atoms C3 and O4.

The corresponding EVB file is automatically generated by the subprogram PREPARE.

3.1.4. The nature of the force field in ENZYMIX

Once the user has defined the quantum atoms and their bonding patterns in each of the resonance forms, the program will automatically compute the parameters of the classical force field that defines the interactions of the atoms in each of the resonance forms. This force field consists of Morse potentials between atoms that are bonded and repulsive potentials between atoms that are not bonded. Also included are potential functions for angles between bonds containing quantum atoms and torsional potentials around quantum bonds. In the above example we will have:

$$\epsilon_1 = M(b_{34}) + \alpha_1 + U^{(1)}_{strain} + U^{(1)}_{electro,repl}(r_{23}) + U^{(1)}_{electro,repl}(r_{24}) + U^{(1)}_{S-s} \tag{2}$$

$$\epsilon_2 = M(b_{34}) + M(b_{23}) + \alpha_2 + U^{(2)}_{strain} + U^{(2)}_{S-s} \tag{3}$$

Where the M(b) terms are Morse potentials between bonded atoms, and the U_{strain} terms are given by:

$$U^{(1)}_{strain} = \frac{1}{2}\sum_{l} K^{(1)}_{b}(b^{(1)}_{l} - b^{l}_{0})^2 + \frac{1}{2}\sum_{m} K^{(1)}_{\theta}(\theta^{(1)}_{m} - \theta^{m}_{0})^2 + \frac{1}{2}K^{(1)}_{\chi}(\chi^{(1)} - \chi_0)^2 \tag{4}$$

$$U^{(2)}_{strain} = \frac{1}{2}\sum_{l} K^{(2)}_{b}(b^{(2)}_{l} - b^{l}_{0})^2 + \frac{1}{2}\sum_{m} K^{(2)}_{\theta}(\theta^{(2)}_{m} - \theta^{m}_{0})^2 \tag{5}$$

where the quadratic bonding terms describe all bonds not represented by the morse potentials and the θ terms represent the angle bending relative to the unstrained equilibrium (120° for the sp^2 hybridization of resonance form I and 109.5° for the sp^3 hybridization of resonance form II). K_χ is the out-of-plane force constant for deforming a planar sp^2 carbon.

$U_{electro-repl}$ represents electrostatic and repulsion interactions between nonbonded quantum atoms and U_{S-s} represents the complicated interaction between the quantum atoms and the remainder of the protein and water atoms (see Ref. 22 for a detailed description of the force field).

The actual PES of the EVB system is computed by a weighted sum of the potentials of resonance forms I and II. The actual potential is given by $\epsilon_{act} = c_1^2\epsilon_1 + c_2^2\epsilon_2 + 2(c_1^2c_2^2)^{\frac{1}{2}} H_{12}$ where the c's are determined by diagonalizing the Hamiltonian:

$$\begin{bmatrix} \epsilon_1 & H_{12} \\ H_{12} & \epsilon_2 \end{bmatrix}$$

and the functions H_{12} are calculated as described in Ref. 22. Note that the matrix elements are all complicated functions of the nuclear coordinates of the protein and must be constantly reevaluated during the course of the dynamics simulation.

3.1.5. Evaluating the reaction profile

Since enzymatic reactions occur on time scales much greater than those accessible by MD simulations (most enzymatic reactions have rates in the order of 1-10^6 $s^{-1}$and MD simulations cannot access times over 10^{-10} s) it is necessary to find some way of *causing* the reaction to happen. The method that we use is the combination of a FEP method and an umbrella sampling method. During each dynamics run the quantum atoms are *constrained* to have a force field with the form $\epsilon_m = (1-\lambda_m)\epsilon_1 + \lambda_m\epsilon_2 - 2((1-\lambda_m) * \lambda_m)^{\frac{1}{2}} H_{12}$. The user does a number of different dynamics runs (11 typically) and for each run the bond order (i.e., percentage of product and reactant character in the reacting atom force field) is held fixed. This means that the Hamiltonian is not diagonalized to get the bond orders. A typical pattern of dynamics runs is as given in Table 1.

Table 1 *Typical pattern of dynamics runs*

Run no.	Reactant bond order	Product bond order	Run-time (in ps)	Purpose
1	1.0	0.0	2.0	equilibration
2	1.0	0.0	1.0	mapping point ($\lambda_m = 0.0$)
3	0.9	0.1	1.0	mapping point ($\lambda_m = 0.1$)
4	0.8	0.2	1.0	mapping point ($\lambda_m = 0.2$)
5	0.7	0.3	1.0	mapping point ($\lambda_m = 0.3$)
6	0.6	0.4	1.0	mapping point ($\lambda_m = 0.4$)
7	0.5	0.5	1.0	mapping point ($\lambda_m = 0.5$)
8	0.4	0.6	1.0	mapping point ($\lambda_m = 0.6$)
9	0.3	0.7	1.0	mapping point ($\lambda_m = 0.7$)
10	0.2	0.8	1.0	mapping point ($\lambda_m = 0.8$)
11	0.1	0.9	1.0	mapping point ($\lambda_m = 0.9$)
12	0.0	1.0	1.0	mapping point ($\lambda_m = 1.0$)

Using the FEP method it is possible to show that the change in free energy of the entire system between two mapping points is given by:

$$\delta G(\lambda_m \Rightarrow \lambda_{m'}) = -\frac{1}{\beta}\ln[<\exp(-(\epsilon_{m'} - \epsilon_m)\beta)>_m] \tag{6}$$

$$\Delta G(\lambda_n) = \Delta G(\lambda_0 \Rightarrow \lambda_n) = \sum_{m=0}^{n-1} \delta G(\lambda_m \Rightarrow \lambda_{m+1}) \tag{7}$$

where $< >_m$ is a configurational average of the quantity in the brackets as the entire system moves on the potential surface defined by ϵ_m. The corresponding free energy profile $\Delta G(\lambda)$ of the reaction in solution is plotted in Fig. 4.

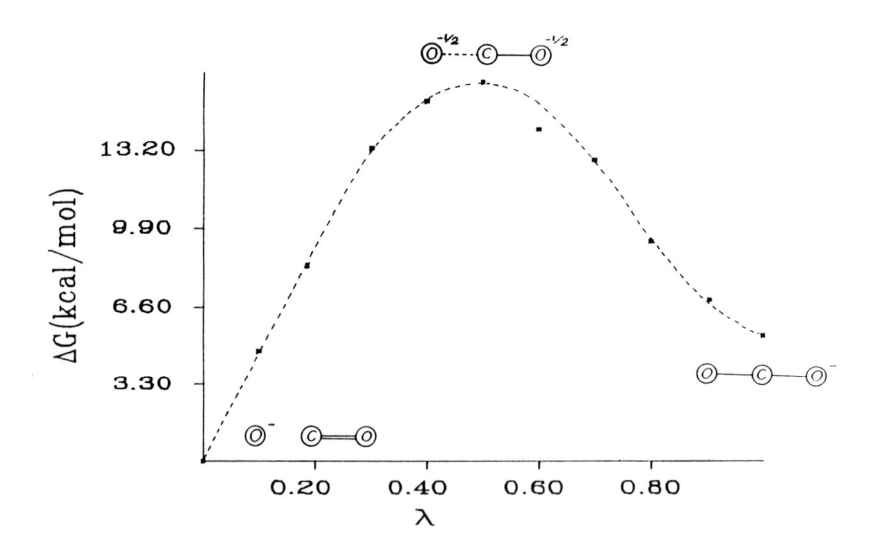

Fig. 4. Free energy of the reaction as a function of the mapping coordinate λ for the $O^{-}+C=O \Rightarrow [O-C-O]^{(-)}$ reaction in solution.

Finally, we can obtain the activation free energy for our reaction $\Delta g^{\ddagger}$ by converting ΔG(λ) to the actual free energy of the EVB ground state surface. This requires a somewhat elaborate treatment which is described in [25] and the result is shown in Fig. 5.

3.1.6. Specialized ENZYMIX *tasks*

In addition to ENZYMIX's normal function of generating a free energy profile for reactions in solution and in a protein it is possible to use the program for a number of specialized tasks of general interest.

(i) Simulation of reactive trajectories. While the major part of enzyme specificity is associated with the corresponding activation barrier (which is evaluated by the EVB + FEP method) one might be interested in dynamical aspects of biological reactions. For example, this can be important in finding out what are the fluctuations of the enzyme which are the most important in inducing the given reaction. A simulation of the actual reactive fluctuation can rarely be accomplished by running dynamics of the enzyme-substrate complex since the chance for the reaction happening during the few tens of picoseconds that a typical dynamics simulation encompasses is extremely small (for a 12-kcal/mol reaction barrier it would require more than 10 years on a Cray to

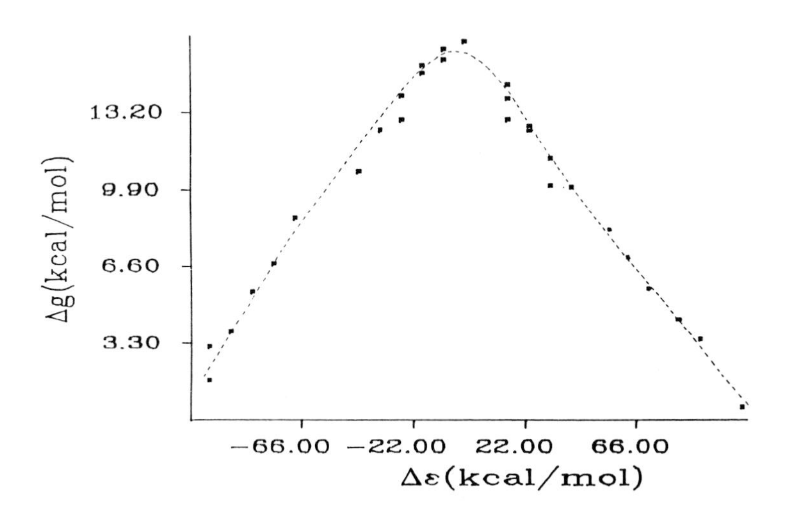

Fig. 5. Free energy surface of the $O^- + C{=}O \Rightarrow [O-C-O]^{(-)}$ reaction plotted as a function of the energy gap ΔE between the reactant and product states.

get just one reactive fluctuation). Fortunately, one can simulate this rare event by starting at the transition state and propagating 'downhill' trajectories. The time reversal of such trajectories gives an ensemble of 'uphill' trajectories that can be studied.

(ii) Calculation of the electrostatic energies of charged groups. The energetics of many biological processes are determined by the free energy of charging the relevant cofactor in the protein site. This includes redox processes, ion binding, ion channels, proton pumps and many other processes (see Ref. 8). In determining electrostatic free energies of charged groups in proteins the key difficulty is not the interaction between charges, but it is in calculating the 'self-energy' or 'solvation energy' (the energy of moving the charged group into it's protein site from vacuum). This energy is best defined relative to the corresponding charging process in water, i.e.:

$$\Delta G^{p} = (\Delta G^{p}_{sol} - \Delta G^{w}_{sol}) + \Delta G^{w} = \Delta\Delta G^{w \Rightarrow p}_{sol} + \Delta G^{w} \qquad (8)$$

where p and w designate water and protein, respectively. With this expression we can formulate conveniently the energetics of various processes. For example, for calculations of absolute pK_a's we use:

$$pK^{p}_{a} = 0.735\Delta\Delta G^{w \Rightarrow p}_{sol} + pK^{w}_{a} \qquad (9)$$

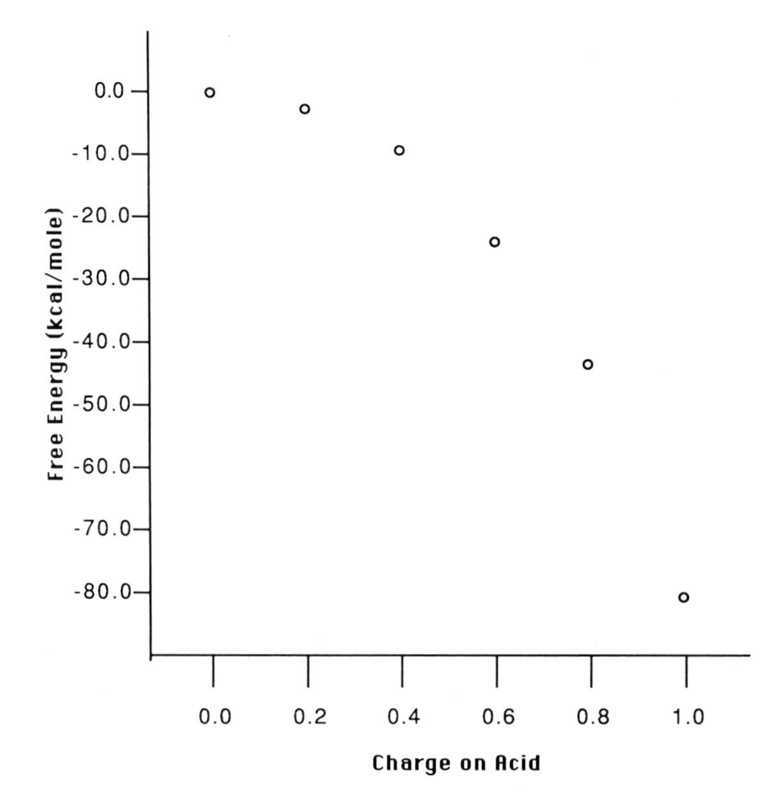

Fig. 6. Calculation of the solvation free energy of an ionized acid in a protein by the adiabatic charging method. The calculations are done for Asp3 in BPTI.

It can be seen that knowledge of the shift in self energies of charged groups in proteins relative to solution is of key importance.

The solvation free energy of the given cofactor in it's protein active site is evaluated by FEP calculations where the atomic charges are changed between the two relevant charge forms (Q_1 and Q_2), i.e.:

$$\epsilon_m = (1-\lambda_m)\epsilon(Q_1) + \lambda_m\epsilon(Q_2) \tag{10}$$

where the two potential surfaces differ only in the charge distribution on the cofactors. Thus this procedure can be considered an 'adiabatic charging process' (see Ref. 26). The typical plot of solvation free energy as a function of λ_m is shown in Fig. 6.

(iii) ENZYMIX calculations of the effect of genetic mutations on catalysis. ENZYMIX can effectively combine the EVB and FEP method in studies of the effect of genetic modifications on the rate of enzymatic reactions [21]. Such studies can be accomplished in two ways. One way is to use ENZYMIX to calculate the reaction profile for the native enzyme and then recalculate the reaction profile for the mutant enzyme (i.e., calculating $\Delta G^{\ddagger}_{nat}$ and $\Delta G^{\ddagger}_{mut}$). The other way is to complete the thermodynamic cycle by mutating the protein at both the transition state and the reactant state thus obtaining $\Delta G^{\ddagger}$s indirectly.

(iv) Calculations of protein-ligand interactions/binding constants. The same approach described above can be used effectively in the evaluation of the effect of mutations on the binding free energies of different ligands and drug molecules. This is done by performing the same calculations as described above but now mutating the ligand/drug from one molecule to the other. The thermodynamic cycle is completed by mutating the ligand/drug in water. The difference between these two free energies gives the difference in the free energy of binding between the two molecules.

4. Electronic Structure of Biological Chromophores and the QCFF/SOL program

Another feature of our simulation approach is the ability to evaluate the properties of medium-sized molecules and biological cofactors in solution and in protein microenvironments. This is done by the QCFF/SOL program which is based on the heteroatom version [27] of the widely used QCFF/PI program with an extremely efficient incorporation of microscopic solvent effects and a direct interface with the electrostatic potential generated by POLARIS or ENZYMIX. This includes the ability to evaluate the solvent effect on quantum mechanical calculations on ground and excited states.

The QCFF/SOL program has been used extensively since the mid 1970s in studies of spectroscopic and conformational properties of molecules in solution and in protein with an emphasis on reliability and efficiency. The program is particularly effective in exploring the actual effect of the protein microenvironment on bound cofactors including such crucial effects as changes in redox potentials, photochemical activity and color control. Also, this program allows the user to study properties of aromatic drugs where quantum mechanical delocalization effects might be significant, to explore solvent effects on electronic and vibrational calculations and to perform molecular mechanics calculations on solvated molecules.

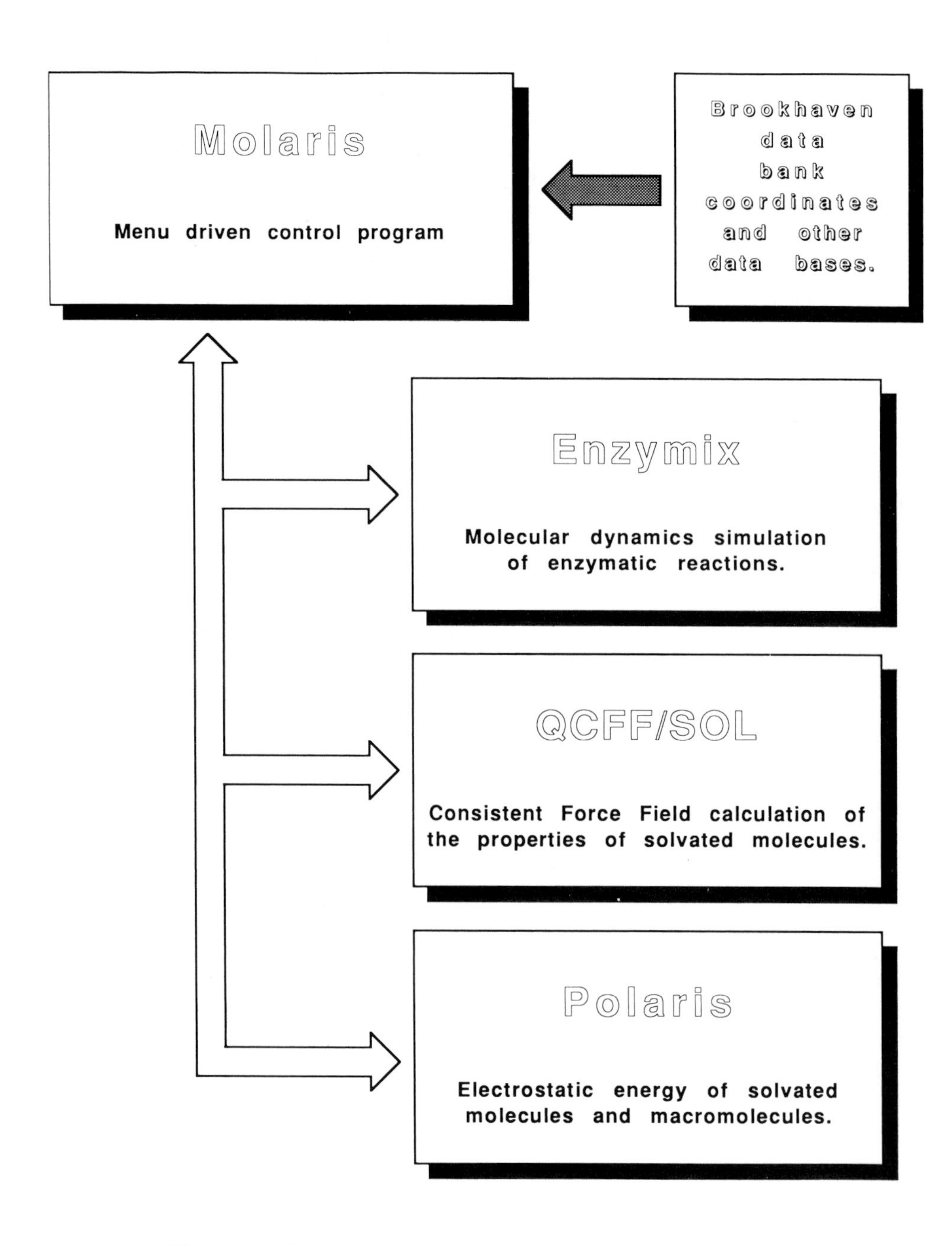

Fig. 7. Relationship of MOLARIS to its various subcomponents.

Features of the QCFF/SOL *program*

The major features of the QCFF/SOL program are:

- Quantum mechanical calculations of solvated conjugated molecules.
- Energy minimization and normal mode analysis of solvated molecules.
- Electronic spectra (UV and CD) of molecules in solution and proteins.
- Solvation free energies of ground and excited electronic states.
- Resonance Raman of biological cofactors in protein sites.
- Charge transfer states in solutions and molecular crystals.
- Photochemical and redox properties of biological factors.
- Conducting polymers.

5. The MOLARIS Package

All of the above simulation tools are integrated in the multi-purpose macromolecular simulation package for computer-aided molecular design called MOLARIS (see Fig. 7 for the structure of the MOLARIS package).

The MOLARIS program is a completely menu-driven program with clear English language commands that provide a convenient interface between the host computer (e.g., the Alliant) and various graphics workstations. This integrated structure is particularly useful in exploring the function of biological macromolecules and exploiting the efficiency and wide range of our simulation methods.

References

1. Warshel, A., Acc. Chem. Res., 14 (1981) 284.
2. (a) Lifson, S. and Warshel, A., J. Chem. Phys., 49 (1968) 5119.
   (b) Warshel, A. and Lifson, S., J. Chem. Phys., 53 (1970) 582.
   (c) Warshel, A., Levitt, M. and Lifson, S., J. Mol. Spectrosc., 33 (1970) 84.
3. Levitt, M. and Lifson, S., J. Mol. Biol., 46 (1969) 269.
4. Born, M., Z. Phys., 1 (1920) 45.
5. Onsager, L., J. Am. Chem. Soc., 58 (1936) 1486.
6. Tanford, C. and Kirkwood, J.G., J. Am. Chem. Soc., 79 (1957) 5333.
7. Warshel, A. and Levitt, M., J. Mol. Biol., 103 (1976) 227.
8. (a) Warshel, A. and Russell, S., Q. Rev. Biophys., 17 (1984) 283.
   (b) Russell, S. and Warshel, A., J. Mol. Biol., 185 (1985) 389.
9. (a) Warshel, A., Biochemistry, 20 (1981) 3167.
   (b) Chung, T. and Warshel, A., Biochemistry, 25 (1985) 1675.
10. Warshel, A., J. Phys. Chem., 86 (1982) 2218.
11. Warwicker, J. and Watson, H.C., J. Mol. Biol., 157 (1982) 671.
12. Zwanzig, R.W., J. Chem. Phys., 22 (1954) 1420.

13. Valleau, J.P. and Torrie, G.M., A Guide to Monte Carlo for Statistical Mechanics: 2. Byways, In Berne, B.J. (Ed.) Modern Theoretical Chemistry, Vol. 5, Plenum, New York, 1977, pp. 169-194.
14. Warshel, A., J. Phys. Chem., 83 (1979) 1640.
15. Warshel, A. and King, G., Chem. Phys. Lett., 121 (1985) 124.
16. Berkowitz, M., Karim, O.A., McCammon, J.A. and Rossky, P.J., Chem. Phys. Lett., 105 (1984) 577.
17. Brooks, C.L. and Karplus, M., Methods Enzymol., 127 (1986) 369.
18. Postma, J.P. and Berendsen, H.J.C., Faraday Symp. Chem. Soc., 17 (1982) 55.
19. Warshel, A., In Chagas, C. and Pullman, B. (Eds.) Pontificiae Academiae Scripta Varia, 55 (1984) 58.
20. Warshel, A. and Sussman, F., Proc. Natl. Acad. Sci. U.S.A., 83 (1986) 3806.
21. Hwang, J.-K. and Warshel, A., Biochemistry, 26 (1987) 2669.
22. Warshel, A., Sussman, F. and Hwang, J.K., J. Mol. Biol., 201 (1988) 139.
23. Wong, C.F. and McCammon, J.A., J. Am. Chem. Soc., 108 (1986) 3830.
24. Singh, U.C., Brown, F.K., Bash, P.A. and Kollman, P.A., J. Am. Chem. Soc., 109 (1987) 1607.
25. Hwang, J.-K., King, G., Creighton, S. and Warshel, A., J. Am. Chem. Soc., 110 (1988) 5297.
26. Warshel, A., Sussman, F. and King, G., Biochemistry, 25 (1986) 8368.
27. Warshel, A. and Lapicerella, A., J. Am. Chem. Soc., 103 (1981) 4664.

Thoughts about the past and future of free energy simulations of biological macromolecules

Jan Hermans
Department of Biochemistry, School of Medicine, University of North Carolina, Chapel Hill, NC 27599-7260, U.S.A.

Introduction

When asked to present a summary at the end of a meeting such as this colloquium, one is likely to end up with a scrambled set of notes, in spite of one's best intentions, as the preceding speakers make the very points one had planned to emphasize. Fortunately, their presentations also produce new material and ideas that can fit in a summary.

The Past

Considerable progress had been made over a number of years by applying the methods of molecular dynamics to biological macromolecules, starting with the work by McCammon, Gelin and Karplus in 1977 on the dynamics of bovine pancreatic trypsin inhibitor [1]. This work emphasized structural and kinetic aspects. Among the former was the question of how closely the mean simulated structure resembled the crystallographic structure used to start the dynamics calculation, the answer being that the resemblance was *not* very good. However, important new insights emerged in the extent to which protein molecules are flexible, and deform as a result of even simple thermal motion. Similarly, the time constants for intramolecular deformations of proteins became for the first time accessible as a result of this work.

A molecular dynamics calculation consists simply of estimating the motion of all atoms of a molecule (to create a trajectory) by solving Newton's equations of motion. These equations relate three variables, i.e., molecular conformation, time and energy. One can order one's thoughts by assigning to each of these three variables one of the three major applications of molecular dynamics calculations: (1) calculations of structure, (2) calculations of kinetics, and (3) calculation of thermodynamic properties. Free energy calculations belong in this last class, and in spite of a late start, have rapidly become the most exciting and productive application of molecular dynamics simulations of proteins.

Applications of molecular dynamics to thermodynamics of biological macromolecules did not get well underway until 1984, when Berendsen and McCammon presented their ideas and first results at a symposium on molecular dynamics of proteins [2] (reviewed by Berendsen and by Beveridge in Ref. 2). There was more than one reason for the delay. Some thermodynamic functions, such as the energy and the specific heat could be extracted very easily from the simulations. However, the entropy, and hence the free energy, could not be estimated directly, and clearly this last function is essential for comparison with the most important class of experimental data, namely, equilibrium constants for molecular interactions and conformation changes. A variety of techniques developed around 1970 by physicists in order to compute free energies of liquids by simulation methods, were found adaptable to problems of protein function and structure, and were in short order applied, by Warshel et al. to the problem of enzyme catalysis [3-5], by McCammon et al. to protein-ligand interactions [6-9], and by Berendsen and coworkers [10, 11] to the interactions of apolar molecules with solvent, particularly water. Recent applications by others have been described in great detail at this colloquium, which attests to the enormous interest in applying simulations to study the thermodynamics of biological macromolecules.

The Present

One of the appeals of free energy simulation is that it appears to work so well; the agreement between calculated and observed quantities is typically called 'astounding'. This astonishment (which includes mine) is not unreasonable; the models by which intermolecular forces are calculated in a typical molecular dynamics simulation have relatively few parameters, seem disarmingly simple, and just do not lead one to expect quantitative predictions. Indeed, agreement between predicted and observed *structures* had been mediocre: root-mean-square deviations of several Ångstroms were obtained in the first calculations, and while these were gradually whittled down to the order of 1 Å, as the models were improved, unconstrained simulated structures remain poor by the experimental standards of protein crystallography. In addition, comparison of computed and observed *kinetic properties* had been able to tell little about the accuracy of the model, because of a paucity of experimental observations on relaxation processes in proteins that occur in the short time span covered by molecular dynamics simulations. On the other hand, agreement between calculated and observed *free energy differences* is, at least in model systems, within a few kcal/mol for differences as large as 15 kcal/mol, and much better for smaller differences; in terms of equilibrium constants, factors of 10^{10} are reproduced within a factor of 100, and factors of 10^5 within a factor of 10. Furthermore, experimental data on the affinity of proteins for small molecules are easy to obtain quite

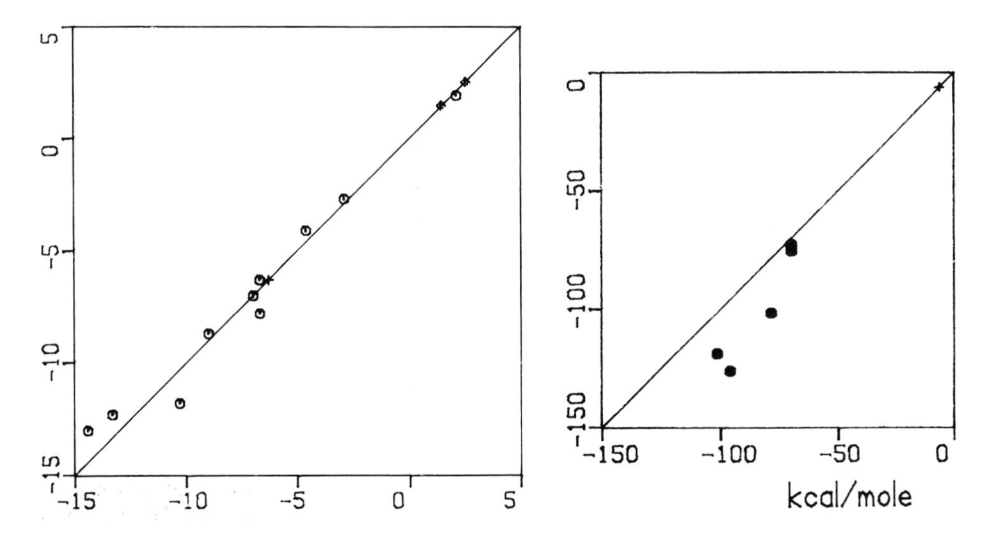

Fig. 1. Correlation of observed and model free energies of transfer of small molecules and inorganic ions from vacuum to water. o, models of amino acid chains (Bash et al. [12 and this colloquium]); ◇, noble gases (Straatsma et al. [11]); +, water (Hermans et al. [14]); •, ions (Straatsma [13]).

accurately, and have traditionally been of great interest to biochemists and pharmacologists.

At this time, a sufficiency of results is available for a first evaluation. I have summarized in two figures several sets of results for problems where the answer is known from experiment. In the first set, available data are collected on the transfer of small molecules from a vacuum to water; these include results for noble gases, small molecules resembling amino acid side chains, and a number of ions. (Fig. 1). The data are from work of Bash and Kollman, presented by Bash at this colloquium, and in Ref. 12, from the work of Straatsma et al. [11, 13], and include a recent result for the transfer of water from liquid to vapor that was computed in our laboratory [14]. The agreement between theory and experiment is indeed good, although the errors become considerable when the magnitude of the free energies is large, as for the ions. The free energy of transferring a water molecule from liquid to vapor calculated for the SPC and TIP models is, within a few tenths of a kcal/mol, equal to what is observed. Since the water models used in these calculations do not reflect the increasing electric polarization and dipole moment of the water molecule as it passes successively from a vacuum or a nonpolar environment, to liquid water, to the neighborhood of an ion, the discrepancies are not unexpected, and are indeed

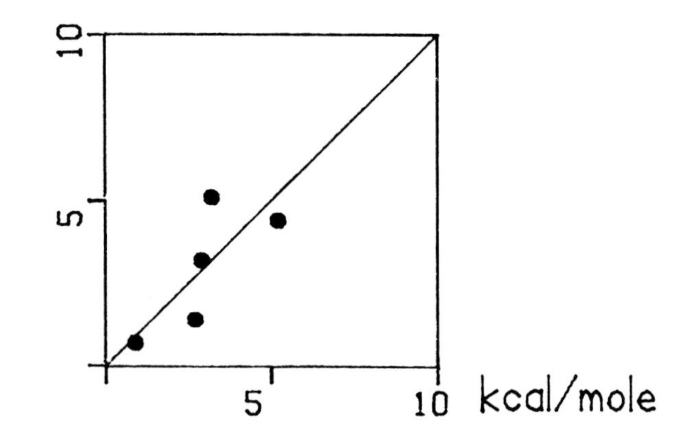

Fig. 2. Correlation of observed and model free energies of binding of methotrexate to mutants of dihydrofolate reductase, from Singh [15 and this colloquium]. The numbers are relative to the wild-type enzyme.

exceptionally small. The same cause, i.e., the failure of the model to adapt its dipole moment to the local electric field, will affect the accuracy of calculated transfer free energies of apolar molecules, but on a much smaller scale.

The second set shows results for a series of dihydrofolate reductase mutants presented by Singh at this symposium (Fig. 2). The quantity studied is the free energy of binding of the inhibitor methotrexate to this series of enzymes. These results are very promising, however the discrepancy from perfect correlation is considerable.

Not inappropriately, it is claimed that free energy simulation is ready to become a tool for predicting differences of free energies of intermolecular interactions. Particularly promising is the potential application of this tool to the development of new enzyme inhibitors as pharmaceuticals. Assuming that the quality of the results obtained by Singh is applicable to other enzyme-inhibitor systems (and scattered results by others lend support to this view), one has the basis for a rational analysis of the potential value of current simulation technology as a tool for predicting changes in affinity. One must consider the cost of doing the simulation and the expected accuracy, versus the required accuracy, the cost of synthesis of new inhibitors, or new mutant enzymes, and the cost of subsequent measurements.

The Future

Application to molecular interactions

Free energy simulations have made sudden and perhaps unexpected demands on computer time, for two reasons. First, one has a seemingly limitless number of technologically and biochemically interesting questions to answer: just think of rationalizing the inhibition constants of all studied inhibitors of any one enzyme and its mutants (the latter now designed and manufactured in the laboratory on order). Second, as the emphasis has shifted from problems of structure and dynamics to problems of equilibrium thermodynamics, there is less reason to analyze the details of most trajectories and conformations. This is because free energy simulations typically pass through a series of artificially constructed intermediates that are physically unrealizable. Thus, each researcher will be able to perform more simulations without being overwhelmed by the time required to analyze the results.

Accordingly, progress in free energy simulations while potentially very rapid, is heavily limited by available computer time. As recently as 5 years ago, the demands of these calculations exceeded the available computer power. At present, each of several research groups is using hundreds of Cray machine hours. In addition, a number of groups have been able to acquire Star array processors, which may have the power of a Cray but at a much lower price. Dedicating one or more array processors full-time to the single task of molecular dynamics is extremely efficient in terms of total cost of hardware and programming. Similarly, the economics of building a hardwired special-purpose machine for molecular dynamics may be justified in terms of the economies of building and operating several copies of the final product; such a machine, called FASTRUN, has been developed at Columbia University [Levinthal et al. in Ref. 2]. Within a short time, we will need a radical increase in computer time to realize possibilities that are now clearly defined. An immediate 10-fold increase appears needed, and does not seem an extravagant objective, given the speed and price of currently available minisupercomputers, and of products under development and nearing announcement. I should like to encourage one or more manufacturers to closely follow the developments of the FASTRUN design, which is about to come on-line; the special feature of this machine is that nonbonded forces are calculated at very high speed by special hardwired boards containing many processing elements that operate both serially and in parallel. It may be possible to transfer this design and construct add-on processors for general-purpose computers, and thereby obtain an enormous speed enhancement at a modest cost.

The pharmaceutical and biotechnology industry can be expected to invest in new computers for rational drug design studies that require simulation capacity in parallel with X-ray determination of physiologically crucial enzymes. The accuracy with which free energy simulations reproduce experimental free energies will be an important factor determining how soon industrial application will begin. It is quite likely that the accuracy of simulations applied to interactions of proteins, as in the work reported by Singh, can be improved without major changes in methodology, essentially at the cost of a higher investment in computer time. At present, these simulations are done under complicating limitations whose only purpose is to reduce the length of the computation. Thus, only a part of the protein, near the active site, is mobile, and water molecules are included only in a shell around the active site. From any standpoint, except that of available computer time, it is easier to work with the entire enzyme molecule mobile, placing it in a box of water molecules, and using periodic boundary conditions as is usual in the study of liquids. This will undoubtedly improve the results of the simulations. It may be particularly important that unless the entire molecule is mobile, global deformations of the protein molecule, some of which correspond to low-energy normal modes, will not be accessible [16]. Yet, precisely these deformations may be essential for the proper accommodation of a series of substrates or inhibitors of different size.

Simulations will also be applied to broad problems of protein-protein and protein-nucleic acid recognition, by a combination of molecular dynamics and site-directed mutagenesis. Looking further ahead, one may expect that our attention will shift to application of free energy simulations to problems of the dynamics of interaction of proteins with other macromolecules, once we have learned how to model equilibria of proteins and inhibitors. This will presumably require as much or more computer time.

Application to protein conformation changes

The success of a free energy simulation may be compromised by the occurrence of a conformation change of the protein or of a radical change in the mode of binding of an inhibitor. As we study more complex problems of enzyme-small molecule interaction, it will be important to develop methods that will take conformational equilibria and alternate modes of binding into account. In fact, the need to model conformational equilibria extends to several important problems, including that of understanding the function of allosteric proteins and the protein folding problem. The latter has proven to be a formidable puzzle, into which insight has been obtained with a variety of ingenious experimental and theoretical approaches, but that remains essentially unsolved. The problem is to find the three-dimensional conformation of a protein when its amino acid sequence is known, without having to go through the difficult and laborious process of X-ray crystallography. Of the two nonexclusive approaches that have promise, one is a heuristic method, in which information about known protein structures, and any experimental data that are informative about the molecular conformation of the test protein, are combined in a search operation [17]. The other method applies an energy criterion, and seeks low-energy conformations [18-20]. These approaches may eventually solve the problem posed by the truly enormous number of possibilities from among which the one correct answer must be found; nevertheless, they are incomplete by themselves, as in the final instance the free energy, and not the energy, determines what is the most stable conformation of a protein. This suggests that free energy simulations that deal with conformational complexity may have an important role in the final stage of a solution to the problem of deducing a protein's conformation from its amino acid sequence.

We have, in our laboratory, begun to systematically explore conformation space of peptides with free energy simulations, with the ultimate aim being its application to problems of structure and function of proteins. At present our experience suggests that these calculations are straightforward but laborious. I discuss here, as an example, the exploration of conformation space of the alanine dipeptide, *N*-acetyl-alanyl-*N*-methylamide [21, 22]. This molecule has two major degrees of freedom, the main-chain dihedrals ϕ and ψ. Past results of energy calculations have shown that the molecule may have several stable conformations, and our exploratory calculations revealed the existence of four, one in each quadrant of the Ramachandran diagram.

In all our simulations we studied one dipeptide molecule in a box of SPC water molecules. As a first step we performed free molecular dynamics calculations. If the simulation was begun with the dipeptide in the extended, β-conformation, then the entire trajectory would explore similar extended conformations. On the other hand, if the calculation was begun with the dipeptide in the right- or left-handed α-conformation, then the trajectory would first explore similar conformations, but eventually cross into the β-region, which indicated that this conformation corresponded to the global free energy minimum. In order to explore the other three quadrants adequately, artificial energy barriers were added in order to retain the conformation inside one quadrant at a time for several hundred picoseconds. Having thereby obtained four independent probability distributions, one in each of the four quadrants, these were put on a common scale by a series of free energy calculations, in which a dihedral angle, ϕ or ψ, was constrained to lie in a narrow interval, and the value to which the angle was constrained was slowly changed during a dynamics calculation. As a result, the molecule would smoothly make the transition from one conformation to another. The free energy change, ΔA_{AB}, for the transition from conformation A to conformation B, was obtained as the total work done by the applied forcing potential in this essentially reversible process with

$$\Delta A_{AB} = \int_{\rho_A}^{\rho_B} T \, d\rho$$

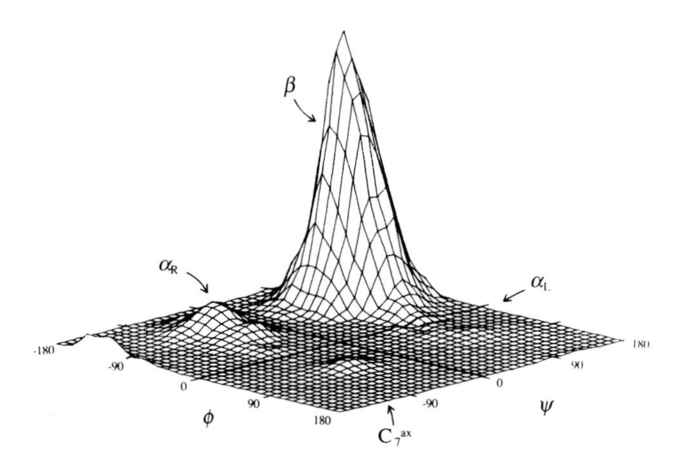

Fig. 3. Probability distribution of the conformation of the alanine dipeptide in aqueous solution from free energy simulation. This map emphasizes the locally stable conformations (probability maxima). From Anderson and Hermans [22].

where T is the applied torque and ρ the angle ϕ or ψ which was forced to change. For completeness, the calculations were done for 8 paths between conformations, each in both directions. Along a cyclic process from one conformation via any other(s), back to the first, the net free energy change should be zero; in practice, the sum of the computed ΔAs was less than a kJ/mol for any such cycle.

Figure 3 shows a map of the probability distribution; the two major conformations are the extended, β-conformation, and the right-handed α-conformation. The former is found to be approximately 10 times more probable than the latter, in agreement with the available experimental observations [23].

Figure 4 shows a free energy map which emphasizes the saddle points between the four conformations found to have local free energy minima. These barriers are low and approximately equal. One concludes that the backbone, at least on either side of a small side chain, can twist readily in all directions during conformation changes and during protein folding.

It has taken us considerably more than a year to complete this project, in part because computer time was not always readily available. We were able to complete the work rapidly thanks to the generosity of the Computer Science Department at Duke University, which placed a Convex-C1 at our disposal with few restrictions. We estimate that the computer time required to now repeat this study with another dipeptide would be of the order of 40 Cray-1 hours. Computer time is clearly a seriously limiting factor in one's choice of problem for which this method is suitable. However, past experience leads us to expect that this limitation will ease significantly in a few years.

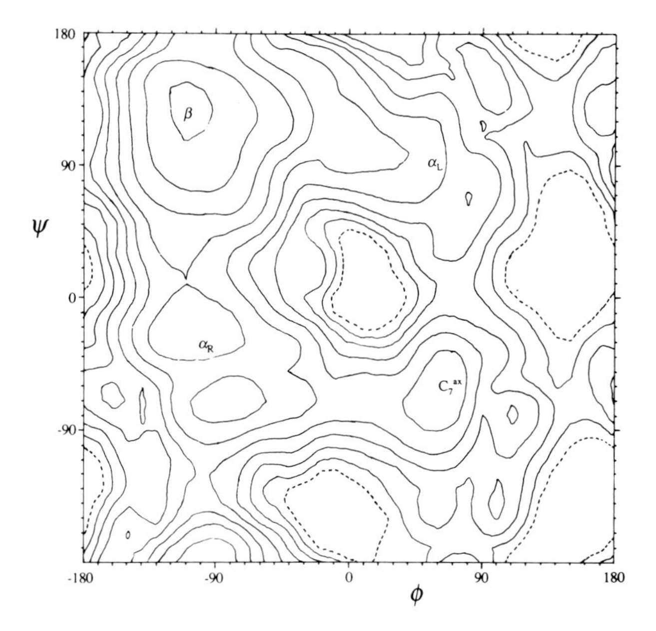

Fig. 4. Conformational free energy map of the alanine dipeptide in aqueous solution with emphasis on the transition regions between the four locally stable conformations. From Anderson and Hermans [22].

Acknowledgements

This work was supported by research grants from the National Science Foundation.

References

1. McCammon, J.A., Gelin, B.R. and Karplus, M., Nature, 267 (1977) 585.
2. Hermans, J. (Ed.) Molecular Dynamics and Protein Structure, Polycrystal Book Service, P.O. Box 27, Western Springs IL 60558, 1985.
3. Warshel A., J. Phys. Chem., 86 (1982) 2218.
4. Warshel A., Sussman, F. and King, G., Biochemistry, 25 (1986) 8368.
5. Hwang, J.-K. and Warshel, A., Biochemistry, 26 (1987) 2669.
6. Tembe, B.L. and McCammon, J.A., Comput. Chem. 8 (1984) 281.
7. McCammon, J.A., Karim, O.A., Lybrand, T.P. and Wong, C.F., Ann. N.Y. Acad. Sci., 482 (1986) 210.
8. Wong, C.F. and McCammon, J.A., J. Am. Chem. Soc., 108 (1986) 3830.
9. Wong, C.F. and McCammon, J.A., Isr. J. Chem., 27 (1987) 217.

10. Postma, J.P.M., Berendsen, H.J.C. and Haak, J.R., Faraday Symp. Soc., 17 (1985) 55.
11. Straatsma, T.P., Berendsen, H.J.C. and Postma, J.P.M., J. Chem. Phys., 85 (1986) 6720.
12. Bash, P.A., Singh, U.C., Langridge, R. and Kollman, P.A., Science, 236 (1987) 564.
13. Straatsma, T.P., Free Energy Evaluation by Molecular Dynamics Simulations, Ph.D. Thesis, University of Groningen, Groningen, 1987.
14. Hermans, J., Pathiaseril, A. and Anderson, A., J. Am. Chem. Soc., 110 (1988) 5982.
15. Singh, U.C., Proc. Natl. Acad. Sci. U.S.A., 85 (1988) 4280.
16. (a) Gō, N., Noguti, T. and Nishikawa, T., Proc. Natl. Acad. Sci. U.S.A., 80 (1983) 3696.
    (b) Brooks, B. and Karplus, M., Proc. Natl. Acad. Sci., U.S.A., 80 (1983) 181.
    (c) Levitt, M., Sander, C. and Stern, P.S., Int. J. Quant. Chem.: Quant. Biol. Symp., 10 (1983) 181.
    (d) Nishikawa, T. and Gō, N., Proteins, 2 (1987) 308.
17. Cohen, F.E., Abarbanel, R.M., Kuntz, I.D. and Fletterick, R.L., Biochemistry, 22 (1983) 4894.
18. Purisima, E.O. and Scheraga, H.A., J. Mol. Biol., 196 (1987) 697.
19. Gibson, K.D. and Sheraga, H.A., J. Comput. Chem., 8 (1987) 826.
20. Li, Z. and Scheraga, H.A., Proc. Natl. Acad. Sci. U.S.A., 84 (1987) 6611.
21. Anderson, A., Carson, M. and Hermans, J., Ann. N.Y. Acad. Sci., 482 (1986) 51.
22. Anderson, A. and Hermans, J., Proteins, 3 (1988) 262.
23. (a) Brant, D.A. and Flory, P.J., J. Am. Chem. Soc., 87 (1965) 663, 2791.
    (b) Madison, V. and Kopple, K.D., J. Am. Chem. Soc., 102 (1980) 4855.

Potential energy functions for organic and biomolecular systems

Arnold T. Hagler*, Jon R. Maple, Thomas S. Thacher, George B. Fitzgerald, and Uri Dinur**

BIOSYM Technologies, Inc., 10065 Barnes Canyon Road, Suite A, San Diego, CA 92121, U.S.A.

Introduction

Theoretical techniques currently being applied to such problems as the structure and fluctuations of proteins, and protein-drug interactions, include molecular dynamics simulations, Monte Carlo simulations, minimizations, and normal mode analysis [1-4]. These techniques ultimately offer the possibility of understanding the complex behavior of peptides, proteins, and nucleic acids in terms of fundamental molecular forces, and thus provide us with an understanding of biomolecules at a level inaccessible to experimental methods alone. All of these theoretical techniques require and depend on the reliability of an analytical representation of the energy surface of the molecules. The energy is given as a sum of simple analytical functions that express the energy required to distort the internal coordinates from some set of standard values, as well as the interactions between nonbonded atoms. The reliability and accuracy of the results of any simulation based on this energy representation depend both on its analytical form and the accuracy to which the force constants have been derived. Thus, it is of crucial importance to derive accurate potential functions, as well as to assess the limitations and deficiencies of the derived functions. Ultimately, this can only be done by calculating a wide range of experimental data and assessing the degree to which the energy surface is capable of reproducing this data.

Background

Here we report preliminary results of a concerted effort to develop and test potential energy functions for organic and biomolecular functional groups. We first review the developments that have brought us to this stage and then describe

*Also at the Agouron Institute, 505 Coast Boulevard South, La Jolla, CA 92037, U.S.A.

**On leave from Department of Chemistry, Ben Gurion University of the Negev, Beer Sheva, Israel a novel method for deriving potential energy functions from ab initio data [5]. The latter makes use of information implicit in the first and second derivatives of the ab initio energy surface with respect to the Cartesian coordinates.

In previous work we have addressed both the derivation of the force constants and the appropriate functional forms used to describe the energy surfaces for biomolecular systems [4, 6-10]. These studies relied for the most part on the fit of experimental properties and followed the Lifson consistent force field approach [11]. In the crystal studies, ab initio calculations were used to obtain information about patterns of charge distribution [10], while the values of the partial charges were determined from experimental properties. In the studies of the energetics of highly-strained molecules, ab initio calculations were used to provide information on the functional form. The coupling between angle bending and bond stretching was shown in the quantum mechanical calculations, and the need for an analytical term in the force field to represent this physical interaction was stressed [8].

The Hydrogen Bond as an Electrostatic and Nonbonded Interaction

Extensive work has been carried out in deriving an intermolecular force field for amides and acids as a part of the derivation of accurate potential functions for peptides and proteins. In one study, we carried out energy minimizations of an extensive set of amide and carboxylic acid crystals with various force fields to assess the various functional forms and to compare their accuracy [9]. The experimental observables in these crystals continue to be the largest experimental data base against which the nature of the hydrogen bond function in amides and the nonbond and electrostatic charges have been tested.

A total of 26 amide and acid crystals were minimized, and a variety of properties were calculated and compared with experiment [9]. The results are summarized in Tables 1 and 2.

The results indicate that overall, to achieve at least this level of fit, no explicit representation of the hydrogen bond is necessary. The MCMS potential [12], which incorporates a 10-12 hydrogen bond expression, does not produce a better fit to experiment in spite of the additional parameters. Thus, it was concluded that the hydrogen bond could be represented as

$$V_{HB} = \frac{A_{ij}}{r_{ij}^{12}} - \frac{C_{ij}}{r_{ij}^{6}} + \frac{q_i q_j}{r_{ij}} \quad (1)$$

Standard nonbonded and electrostatic interactions are employed *with only one exception: the nonbonded parameters of the polar hydrogen atom in the hydrogen bond are zero*; (i.e., $A_{HN} = A_{HO}$, $C_{HN} = C_{HO} = 0$, where A_{HN}, A_{HO} are the repulsive coefficients and C_{HN}, C_{HO} are the dispersive coefficients of the hydrogen on the nitrogen and oxygen). The benchmark also pointed out apparent common deficiencies in all functional forms. For example, crystals such as acetic acid, α-oxalic acid, and glutaramide, which were fit poorly by one force field, were generally problematic in the other two force fields as well [9]. This is, perhaps, the most valuable information we can obtain from this type of study because it provides key, well-defined observables to test proposed improvements in the potential function.

Examination of spatial electron densities provides some insight as to why the polar hydrogen radius goes to zero in fits of crystal properties [13]. The size and shape of the molecule (or its component groups) are directly related to the volume from which other molecules, or other parts of the same molecule, are excluded. This factor, which in empirical conformational calculations is

Table 1 *Root mean square deviations of properties calculated for carboxylic acids and amides by various force fields*

Property	Units	No. of observables	Root mean square deviations		
			12-6-1	9-6-1	MCMS
Acids					
Energy	kcal/mol	12	2.486	2.053	2.118
UCV length	Å	42	0.489	0.307	0.604
UCV angle	deg	17	3.456	2.856	4.465
Volume	$Å^3$	14	15.911	16.772	18.876
d<4	Å	14	0.247	0.190	0.322
H-O dist	Å	16	0.062	0.072	0.058
O-O dist	Å	16	0.047	0.071	0.041
C-O-O angle	deg	16	11.071	9.881	14.048
O-O-C angle	deg	16	7.843	7.760	11.786
H-O-C angle	deg	16	12.362	12.144	17.985
180° - O-H-O	deg	16	8.491	7.732	11.710
Amides					
Energy	kcal/mol	6	1.574	1.930	8.446
UCV length	Å	36	0.208	0.235	0.261
UCV angle	deg	14	1.824	1.261	2.385
Volume	$Å^3$	12	7.057	17.797	13.951
d<4	Å	12	0.145	0.145	0.164
H-O dist	Å	30	0.049	0.059	0.056
N-O dist	Å	30	0.055	0.055	0.076
C-N-O angle	deg	22	3.337	3.575	4.071
N-O-C angle	deg	22	5.931	5.502	9.257
H-O-C angle	deg	30	5.830	5.609	7.329
180° - N-H-O	deg	30	4.396	3.894	4.093

Table 2 *Root mean square deviations of the lengths (Å) and angles (deg) of the unit cell vectors*

Molecule	9-6-1		12-6-1		MCMS	
	Lengths	Angles	Lengths	Angles	Lengths	Angles
Acids						
Formic acid	0.22	0.0	0.16	0.0	0.50	0.0
Acetic acid	0.62	0.0	0.97	0.0	1.24	0.0
Propionic acid	0.09	0.5	0.17	0.3	0.28	2.0
Butyric acid	0.43	5.8	0.25	5.3	0.27	7.0
Valeric acid	0.18	0.7	0.20	1.4	0.18	0.6
α-Oxalic acid	0.70	0.0	1.11	0.0	1.39	0.0
β-Oxalic acid	0.23	1.9	0.83	8.1	0.84	9.7
Malonic acid	0.15	2.9	0.13	3.2	0.14	4.6
Methylmalonic acid	0.14	1.6	0.20	1.7	0.21	2.2
Succinic acid	0.12	0.3	0.18	1.2	0.13	1.4
Glutaric acid	0.20	0.4	0.32	1.7	0.39	0.7
Adipic acid	0.09	0.4	0.30	1.1	0.43	2.3
Suberic acid	0.11	0.1	0.15	2.5	0.20	3.0
Sebacic acid	0.09	0.9	0.19	0.2	0.20	0.1
Amides						
Oxamide	0.03	1.1	0.16	3.2	0.10	4.4
Malonamide	0.10	0.3	0.11	0.8	0.27	0.2
Succinamide	0.05	1.0	0.07	0.1	0.11	2.5
Glutaramide	0.49	2.5	0.47	2.5	0.44	1.8
Adipamide	0.12	0.2	0.06	1.0	0.05	0.7
Urea	0.09	0.0	0.07	0.0		
Formamide	0.40	1.2	0.36	2.1	0.45	0.5
Diketopiperazine	0.09	1.4	0.10	0.2	0.11	2.5
LL-Dimethyl-diketopiperazine	0.27	1.4	0.21	1.1	0.24	1.2
Cyclopropane-carboxamide	0.29	1.0	0.10	0.5	0.31	0.1
N-Methylacetamide	0.18	0.0	0.20	0.0	0.30	0.0
Suberamide	0.17	0.8	0.09	0.6	0.03	1.0

represented by the nonbonded repulsion, is one of the important factors in determining the conformation of large molecules [14, 15] (and crystal packing modes [6, 16]). The representation of the nonbonded atom-atom repulsion is one example where the implicit assumption of spherical atoms is introduced in empirical conformational analysis. The shapes of the calculated electron densities, along with the analysis of the contour lines, indicate that for the repulsive nonbonded potential, this widely-used spherical atom approximation seems to be a reasonable first approximation in representing this important excluded volume effect [17, 18]. In fact, the small distortion of atoms in a molecule is a necessary condition for obtaining precise results from determinations of molecular structure from X-ray crystallography where, in general, the scattering is assumed to take place from spherical atoms [17]. Correlated to this is the difficulty encountered in determining the small migration of electrons on molecular formation by these techniques [18, 19].

A description of the overall shape of the molecule may be obtained by drawing a surface of constant electron density in space, which envelopes the molecule. Such plots for the amide group in AcNHMe, and the carboxylic acid group of acetic acid, are presented in Figs. 1 and 2, respectively. Of course, the resulting shape depends to some extent on the density level chosen. In Figs. 1A and 2A, the contour level chosen is 0.027 electrons/$Å^3$, which corresponds to a surface of constant density at a distance from the atomic centers that is approximately the van der Waals radii of the atoms [12, 20]. In Figs. 1B and 2B, the level of 0.75 electrons/$Å^3$ [21], used by Dunning and Winter [22], is presented for comparison.

These three-dimensional shape plots verify qualitatively the conclusion reached by observing the contour plots. The overall shapes of the atoms appear spherical at both contour levels. The shape of the molecule emerging from the van der Waals surface is quite diffuse and to some extent it loses the features of distinct atoms that is seen in the 0.75 electron/$Å^3$ [21] shape plot. In particular, it can be seen that at the van der Waals radius the hydrogens in the amide (and to a slightly lesser extent in the hydroxyl group) are enveloped by the nitrogen and oxygen density. To an external probe, this density appears to be a single spherical cloud of electrons surrounding the nitrogen (or oxygen). This explains the negligible van der Waals parameters for the hydrogen. Of course, the proton and partial positive charge are still located beneath this electron cloud (see the difference density maps [13]); thus, we continue to have a positive partial charge at this hydrogen that plays an important role in the energetics.

Thus, from the discussion above we see the interplay of one aspect of ab initio molecular orbital calculations with crystal data. Other applications of crystal data to potential function development have been carried out by Momany et al. [23] and Williams and Starr [24] and have been reviewed earlier [4]. Such crystals provide a powerful test of van der Waals and electrostatic parameters and more advantage could be taken of this data when an existing, currently used functional form is tested. The basic premise is that potential functions should be able to account for the structure (lattice constants) and sublimation energies of a large data base of simple model compounds such as amides, acids [9], etc., if they are to provide accurate information on the more complex questions involving protein structure, ligand binding energies, and free energies.

Similarly, experimental data on isolated molecules, including structures obtained from electron diffraction studies, rotational barriers, and strain energies, as well as vibrational spectra (which is, perhaps, the most powerful data), have

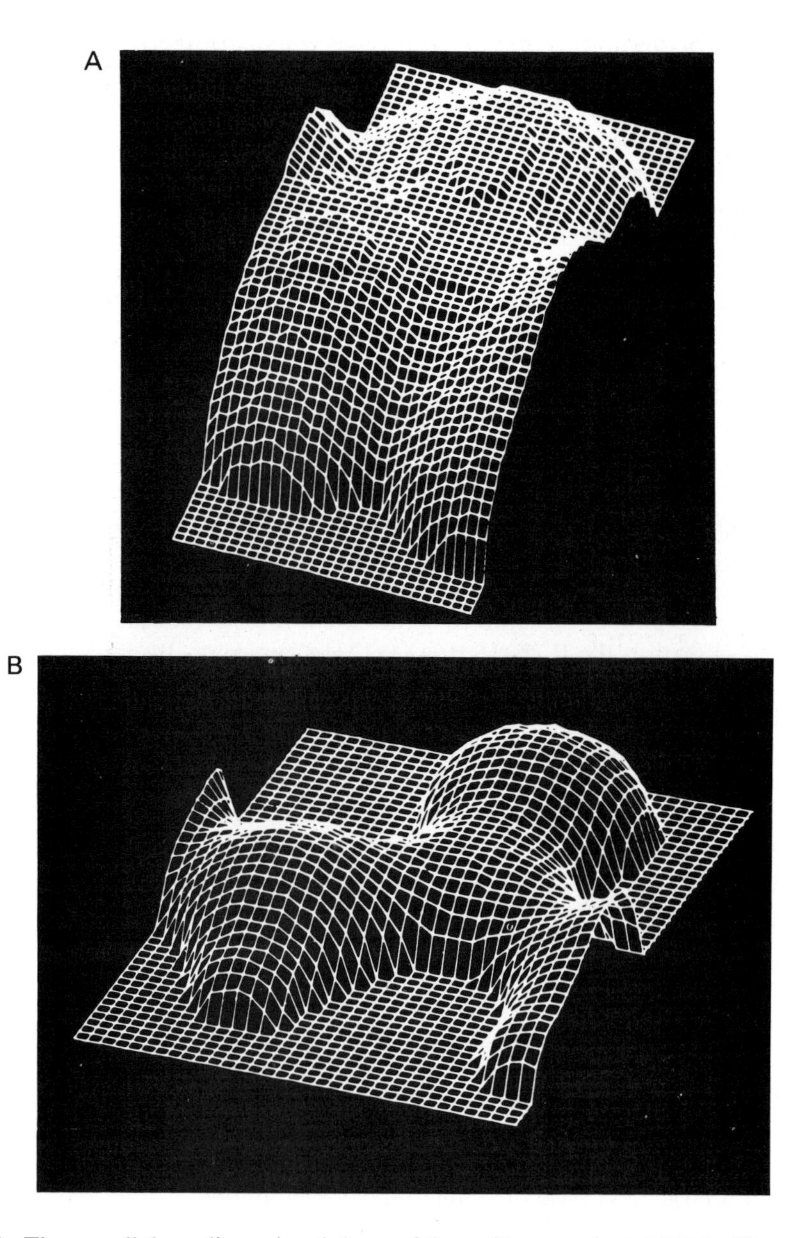

Fig. 1. The overall three-dimensional shape of the amide group in AcNHMe. The carbonyl group is on the right with the oxygen at the top of the figure, while the N-H is trans-directed toward the lower border of the plane. (A) Surface of constant electron density of 0.027 electrons/Å^3, roughly corresponding to van der Waals radius. (B) Surface of constant electron density of 0.75 electron/Å^3.

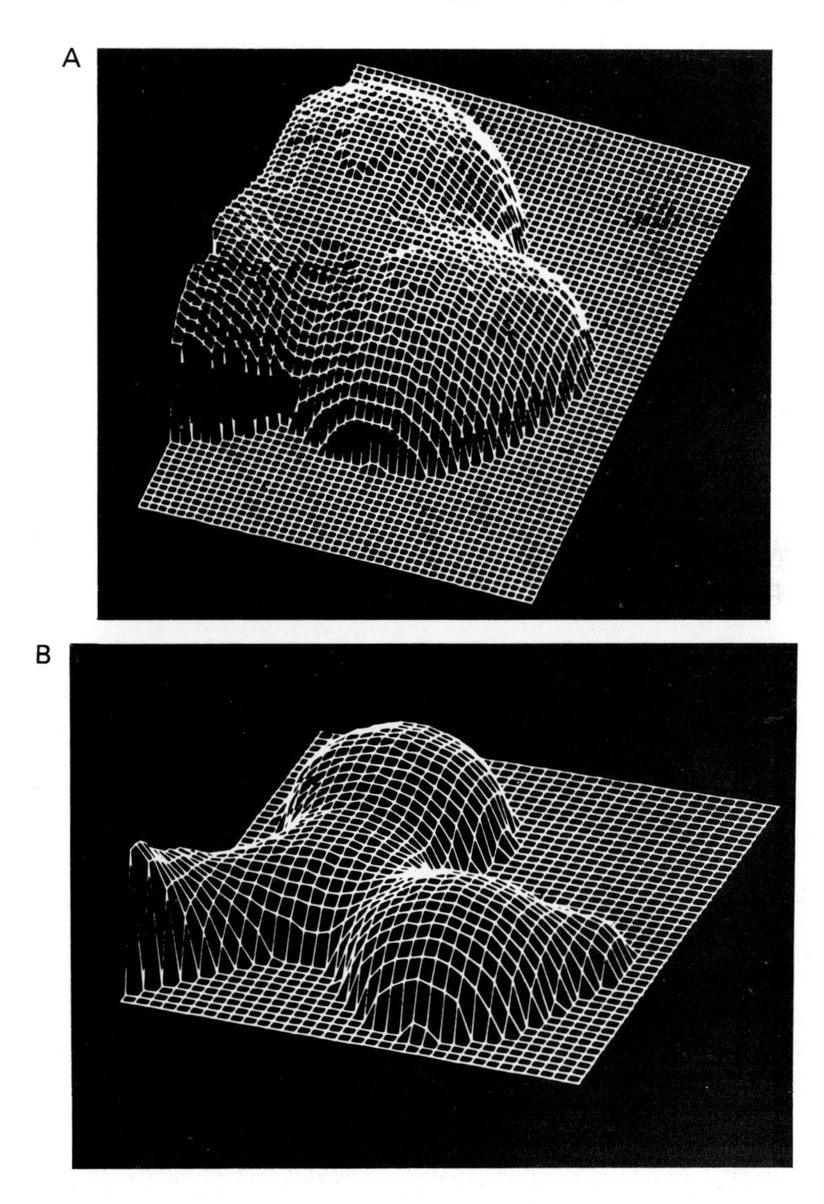

Fig. 2 The overall three-dimensional shape of the carboxyl group in acetic acid. The carboxyl group is on the left with the oxygen at the top of the figure, while the O-H is cis and in the foreground directed to the right. (A) Surface of constant electron density of 0.027 electrons/Å^3, roughly corresponding to van der Waals radius. (B) Surface of constant electron density of 0.75 electrons/Å^3.

been used to obtain intramolecular potential functions. An example of one of the most recent applications is that of Lifson and Stern [25], which is the latest work in the development of a force field for hydrocarbons. Table 3, which

Table 3 *Root mean square deviation of calculated from observed[a] vibrational frequencies (in cm⁻¹)*

Alkane molecule[b]	CH Stretch		Others		Total	
	LS[c]	WL[d]	LS	WL	LS	WL
Methane	9.2		3.5		7.0	
Ethane[e]	28.2	20.1	7.7	32.4	17.4	28.9
	12.8	5.7			8.9	29.1
Propane	10.6		10.5		10.6	
n-Butane	6.6	15.1	6.3	26.5	6.4	23.1
Isobutane	11.0	26.9	7.5	32.1	8.8	30.5
Neopentane	18.5		22.2		20.7	
Cyclopentane[f]	13.0	23.6				
Cyclohexane[g]	16.0	17.5	16.7	25.7	16.5	24.2
TTBM[f]	8.0	41.6				

[a] See Refs. 42-45.
[b] Only the first five molecules were included in the optimization.
[c] This work.
[d] The CFF of Ref. 37, except as noted.
[e] Upper values include symmetric CH_3 stretching frequencies, while lower values exclude them; see the text for discussion.
[f] Because of the lack of experimental assignments, only the CH stretch is listed. WL for cyclopentane is from Ref. 36; for TTBM as in Ref. 8.
[g] For consistency, the same observed frequencies were used as in Ref. 36.

summarizes their results, shows that an excellent fit to experiment has been achieved with the mean errors in frequencies of 10-20 cm^{-1}. This sort of accuracy ultimately will be needed in all functional groups if we are to rigorously predict biomolecular dynamic and thermodynamic behavior.

The approach outlined above is basically the consistent force field approach pioneered by Lifson. Lifson's foremost goal was to derive force fields that are transferable and generally applicable over the range of intra- and intermolecular deformations sampled in standard applications. These are crucial requirements if they are to be continuously applied to new systems with confidence. His premise was that to achieve these requirements, the same force field must be able to fit a wide range of experimental properties, including molecular structure, vibrational spectra, crystal structure, sublimation energy, and strain energy. In other words, the energy function must reproduce structure, energy, and dynamics (i.e., vibrational spectra), which are of interest today. Another way of looking at this requirement is that the potential function must have the correct energy, slope or first derivatives (i.e., the molecular or crystal structure), and curvature or second derivatives (i.e., the vibrational frequencies). It is clear that if these properties are not reproduced, it is unlikely that the functional form can be extrapolated beyond the region it was fit to. This is a problem if one wants to use it to predict new systems. We have initiated a major effort to continue and extend this approach to the derivation of rigorous potential energy functions for applications to biomolecular and organic systems. To this end, we have undertaken the compilation and fit of a wide range of experimental properties of an array of functional groups [26]. This has resulted in a data base that now includes experimental properties of 500 compounds. We have also extended the approach to take advantage of the powerful new information which is available from ab initio molecular orbital calculations which are made possible by algorithms for analytical first and second derivatives, as well as the new generation of powerful computers. The use of ab initio *and* experimental data can provide a powerful methodology for addressing questions of functional form, transferability, and providing efficient methods for rapidly obtaining objective sets of force constants.

Force Fields from ab initio Energy Surfaces [5]

This technique has been developed with several objectives in mind. First, it provides an efficient method for obtaining a set of potential constants for an arbitrary functional group of interest from an ab initio surface. Thus, it provides well-defined and well-characterized approximate force constants for functional groups for which no force field is available. More importantly, it may allow us to address two of the most important issues outstanding in the derivation of force fields for biomolecular and organic systems: the validity of alternative analytical functional forms for representing the energy surface of these molecules, and the transferability of both functional forms and potential constants from model compounds to the molecules of interest.

Our approach in the use of ab initio deviates from existing methods in two respects. First, instead of generating and fitting a set of energies by, for example, rotating about a given bond or generating a grid of the potential energy surface [27], we use an analytical evaluation of the first and second derivatives [28]. By fitting the first *and second derivatives* of the energies for a given analytical function to the corresponding values obtained from ab initio calculations on a small number of *distorted* configurations, we are able to obtain enough information about the potential energy surface in the sampled region, including its anharmonic features, to determine the corresponding force constants. The degree to which the surface is fit allows us to judge the adequacy of a particular functional form and to compare a variety of analytical representations.

The Method

One of the objectives of this methodology, as noted above, is to find the analytical form that best describes a given ab initio potential surface. That is, we wish to determine both the functional form (a trivial example would be to assess the ability of a quadratic term or Morse potential to describe bond distortion), and the force constants for each interaction term. To accomplish this in a rigorous and efficient manner, we make use not only of the quantum mechanical energy, but of its first and second derivatives as well. The power of the method lies in the amount of information about the ab initio potential surface that is contained in the derivatives.

For a molecule with n atoms there are 3n first derivatives and $3n(3n+1)/2$ second derivatives, or, excluding rotation and translations, $(4.5)(n-2)(n-1)$ total internal first and second derivatives for each molecular configuration. This constitutes a sizable amount of data, and a relatively small number of properly chosen configurations can provide ample information about the potential surface surrounding the equilibrium geometry. For example, five configurations of a six-atom molecule result in 454 data points (including the four relative energies), and this should suffice for determining a good quality force field with 50 parameters. As noted by Pople et al. [28], the evaluation of both first and second derivatives requires four to five times more computation time than the evaluation of the gradients, and seven times more for a simple Hartree-Fock (HF) calculation. It follows that, with the same amount of computation time, the number of data points that can be obtained for a six-atom molecule is 13 times greater when first and second derivatives are used than when simply calculating the energy at various points. This ratio increases for larger molecules and has a qualitative effect on the ability to derive and characterize force fields.

An efficient manner for obtaining the distorted configurations is to displace the atoms along a linear combination of the vibrational normal modes. By definition, these are orthogonal; therefore, one can be assured of sampling all internal distortions in the molecule. In some applications (e.g., the study of rotational barriers) the coordinates of interest are systematically varied, while the other coordinates are distorted at random.

Once the configurations are determined, a set of ab initio calculations is carried out to determine the energy and its first and second (Cartesian) derivatives for each of them. The task then is to fit these derivatives with an analytical expression of the potential energy (i.e., empirical potential function). As noted above, the ab initio data includes

relative energies ΔE_a^o

$$\text{energy gradients } \frac{\partial E_a^o}{\partial x_i} \equiv g^o_{a,i}$$

$$\text{energy second derivatives } \frac{\partial^2 E_a^o}{\partial x_i \partial x_j} \equiv H^o_{a,ij}$$

We denote the corresponding quantities derived from the empirical potential energy function by ΔE_a, $g_{a,i}$ and $H_{a,ij}$. These will, of course, depend on the parameters (force constants) that appear in the empirical function. We then form the sum of squared deviations

$$S_g^2 = \sum_{a,i} (g_{a,i} - g^o_{a,i})^2$$

$$S_H^2 = \sum_{a,ij} (H_{a,ij} - H^o_{a,ij})^2$$

$$S^2 = W_g S_g^2 + W_H S_H^2 \tag{2}$$

where W_g and W_H are weighting factors. Minimizing S^2 with respect to the potential constants results in the best fit of the given analytical form (or empirical potential function) to the ab initio surface. It is clear that by fitting various analytical forms in this way, one can determine which forms best reproduce the ab initio energy surface; i.e., the ab initio energy as a function of the distortion of the molecule. By fitting ab initio energy surfaces of different molecules with similar functional groups, one can assess the transferability of the potential functions.

An Example

Here we show how an ab initio energy surface can be used to determine the adequacy of various representations of the potential energy as a function of the molecular coordinates. For this purpose we test the adequacy of several levels of approximation in the analytical representation of the force field to account for the energy surface of formate ion, a small model compound. In this case, eight different configurations of this molecular ion were generated by performing a random walk around a reference point estimated to be near the (a priori unknown) equilibrium geometry (see Fig. 3) [5].

Ab initio 4-31G* Hartree-Fock calculations of the energies and their first and second derivatives were then performed for all eight structures, yielding

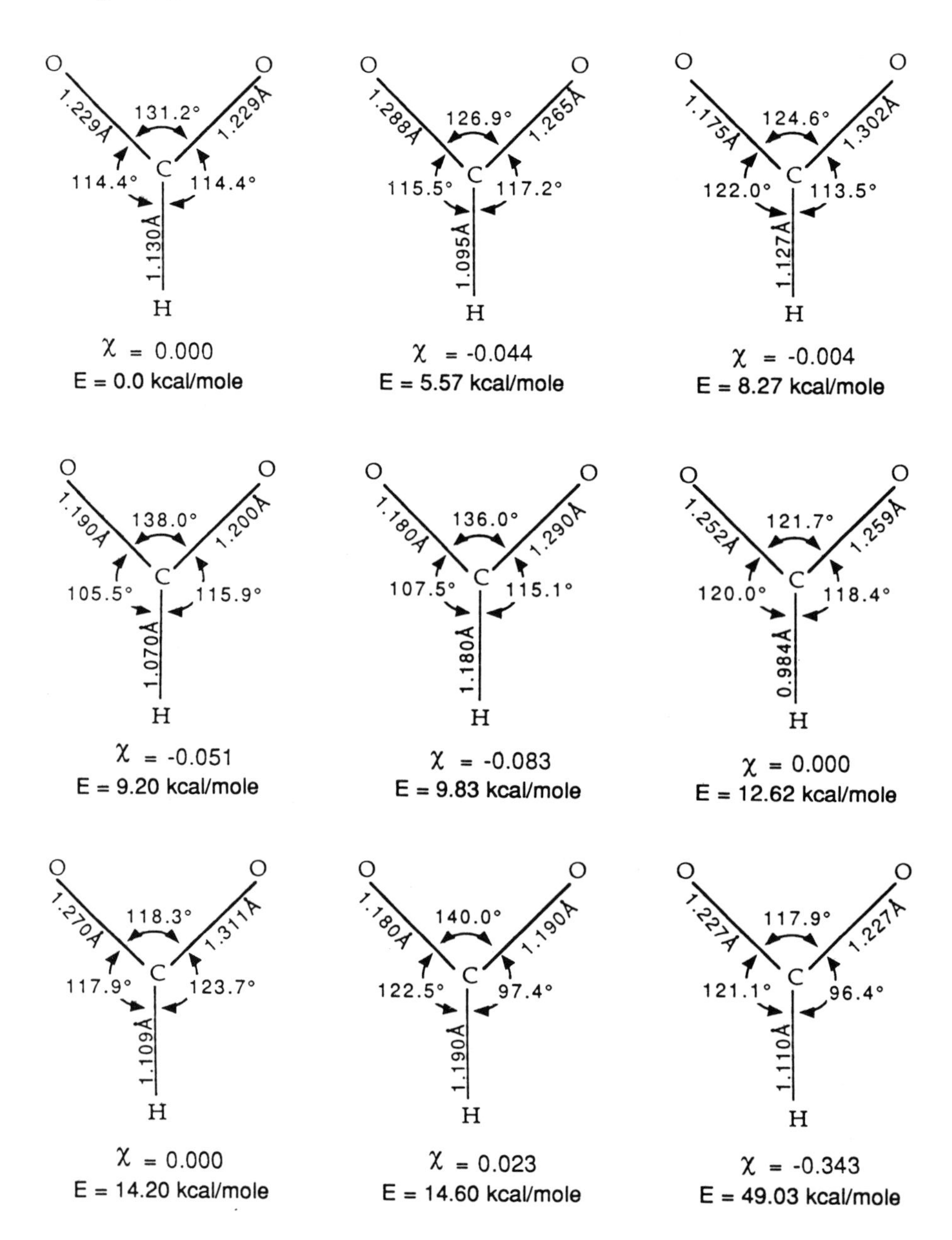

Fig. 3. Configurations of formate anion used to generate the ab initio energy surface. Top left: Equilibrium configuration (E=0.0 kcal/mol).

a total of 720 Cartesian first and second derivatives and seven relative energies. The ab initio energy was subsequently minimized to obtain the equilibrium geometry of the molecule, its energy, second derivatives, and the harmonic spectrum. The energies of the distorted configurations spanned a range of 49 kcal/mol above the equilibrium point, much of which was stored in the out-of-plane mode. From all these data, only the first and second derivatives of the energy of the eight distorted configurations were used to derive the potential constants. The rest of the data (i.e., relative energies, equilibrium geometries, and harmonic spectra), were used to test how well the optimized force fields could predict observables outside the domain of the fitting.

In this work we were interested in testing the ability of the method to fit the ab initio energy surface and to identify those functional forms most appropriate for representing the molecular potential surface. Accordingly, we have performed a series of calculations in which the ab initio data were fitted with a hierarchy of potential functions that differ with respect to the degree of elaboration as well as the system of coordinates that is being used.

We begin with a conventional system of 'springs', which consists of three bond deformations b (C-H and two C-O bonds), three angle deformations Θ (O-C-O, and two O-C-H), and an out-of-plane coordinate χ. The latter is not uniquely defined, and several possibilities exist in the literature. For the following force fields, we have defined this coordinate as the distance of the central atom from the plane formed by the other three. We have also addressed the question of which functional form best represents the energy of this deformation as a function of coordinates.

Six force fields were tested with this system of 'springs'. The first one (1) was a simple diagonal force field in which the molecule was represented by seven noninteracting harmonic springs. Each spring was characterized by a coordinate s (bond, angle, or out-of-plane coordinate) and two parameters - a force constant k and a 'strain-free' reference value s_0. Basically, this simple diagonal form is the form of the force field used in most protein simulations [29-34]. The second force field (2) was diagonal, but included anharmonicity in the form of cubic terms for all springs except the out-of-plane, which is bound by symmetry to include only even terms. The third force field (3) was also diagonal, but the anharmonicity was represented by quartic terms (for all springs) in addition to the cubic ones. The next three force fields allowed the springs to interact. Force field 4 involved the addition of quadratic cross terms to the diagonal harmonic force field (1). Except for the lack of anharmonicity in the bonds, this is essentially the form of the force field used by Lifson for alkanes and Hagler and coworkers for peptides and proteins [1, 8, 25, 35-38]. Force field 5 had quadratic and cubic cross terms added to the cubic force field (2). Similarly, force field 6 resulted from the addition of quadratic and cubic cross terms to the fully anharmonic quartic force field (3). The number of parameters for these three force fields was greatly reduced by excluding coupling terms that are symmetry forbidden or that are redundant because of the redundancy in the coordinates [27, 39, 40]. The constants in these force fields were fit to the ab initio energy surface (i.e., the 720 first and second derivatives of the configuration shown in Fig. 3), by the procedure described above.

Diagonal Force Fields

$$\text{ff 1: } V_1 = \sum_s k_s^{(2)} (s - s_o)^2 = \sum_b K_b (b - b_o)^2 + \sum_\theta K_\theta (\theta - \theta_o)^2 + \sum_\chi K_\chi \chi^2$$

$$\text{ff 2: } V_2 = \sum_s [k_s^{(2)} (s - s_o)^2 + k_s^{(3)} (s - s_o)^3]$$

$$\text{ff 3: } V_3 = \sum_s [k_s^{(2)} (s - s_o)^2 + k_s^{(3)} (s - s_o)^3 + k_s^{(4)} (s - s_o)^4]$$

With Cross Terms

$$\text{ff 4: } V_4 = \sum_s k_s^{(2)} (s - s_o)^2 + \sum_{ss'} k_{ss'}^{(11)} (s - s_o)(s' - s'_o)$$

$$= \sum_b K_b (b - b_o)^2 + \sum_\theta K_\theta (\theta - \theta_o)^2 + K_\chi \chi^2 + \sum_{bb'} K_{bb'} (b - b_o)(b' - b'_o) +$$

$$\sum_{b\theta} K_{b\theta} (b - b_o)(\theta - \theta_o)$$

$$\text{ff 5: } V_5 = \sum_s [k_s^{(2)} (s - s_o)^2 + k_s^{(3)} (s - s_o)^3] +$$

$$\sum_{ss'} [k_{ss'}^{(11)} (s - s_o)(s' - s_o') + k_{ss'}^{(21)} (s - s_o)^2 (s' - s_o') + k_{ss'}^{(12)} (s - s_o)(s' - s_o')^2] +$$

$$\sum_{ss's''} k_{ss's''}^{(111)} (s - s_o)(s' - s_o')(s'' - s_o'')$$

$$\text{ff 6: } V_6 = \sum_s [k_s^{(2)} (s-s_o)^2 + k_s^{(3)} (s-s_o)^3 + k_s^{(4)} (s-s_o)^4] +$$

$$\sum_{ss'} [k_{ss'}^{(11)} (s-s_o)(s'-s_o') + k_{ss'}^{(21)} (s-s_o)^2 (s'-s_o') + k_{ss'}^{(12)} (s-s_o)(s'-s_o')^2] +$$

$$\sum_{ss's''} k_{ss's''}^{(111)} (s-s_o)(s'-s_o')(s''-s_o'')$$

Importance of Cross Terms

The relative rms deviations between the ab initio derivatives of the eight distorted configurations and the corresponding derivatives that were obtained from the optimized empirical force fields are given in column 4 of Table 4 [5]. As expected, more elaborate force fields fit the ab initio data progressively better, but some modifications are more effective. The two most significant modifications are the addition of cubic terms to the harmonic diagonal force field, which leads to a reduction of more than a factor of four in the sum of squares of residuals, and the further addition of cross terms to the cubic diagonal force field, which reduces the sum of squares by an additional factor of 20. Other modifications have a much smaller effect on the sum of squares. In particular, there seems to be an intrinsic limit to the fit that can be obtained with a purely diagonal force field (e.g., force fields 2 and 3).

Although the force fields in this work are derived by fitting energy derivatives, their performance with respect to observables such as structure, energy, and dynamics (as reflected in vibrational frequencies) is of more direct interest. Columns 5-8 of Table 4 display the deviation of the empirical calculations from the ab initio values with respect to the relative energies of the eight distorted configurations, the equilibrium geometry, and the harmonic spectrum. *None of these observables was included in the fitting procedure*, and consequently, the magnitude of the deviations directly demonstrates the predictive power of the force fields. By inspecting the predicted energies (column 5), one sees that inclusion of anharmonicity (through cubic and quartic terms) in a diagonal force field improves the fit of the derivatives, but does not necessarily yield more accurate relative energies. In addition, the anharmonic diagonal force fields are not significantly better than the diagonal harmonic force field with respect to the equilibrium geometry (columns 6 and 7), but are definitely worse with respect to the harmonic frequencies. The latter finding merely indicates that anharmonicity cannot replace the cross terms. This is further manifested in the fact that the results are substantially and uniformly improved when cross terms are included in the force field (see force fields 4-6). For example, upon going from

Table 4 *Formate calculations*

Co-ordinates	Force field	Parameters	Distorted configurations		Equilibrium geometry		
			$\sqrt{S/S_o}$	ΔE rms[a]	Δr rms[a]	$\Delta\theta$ rms[a]	$\lvert\Delta\nu\rvert$ avg[a]
3r, 3θ,χ							
	Diagonal						
	1. quadratic	9	0.298	2.18	0.017	0.27	155.6
	2. cubic	13	0.137	2.99	0.010	0.41	186.6
	3. quartic	18	0.125	2.05	0.009	0.62	172.3
	With Cross Terms						
	4. full 2nd order	14	0.278	1.00	0.010	0.15	88.3
	5. full 3rd order	35	0.032	0.65	0.002	0.42	31.1
	6. full 3rd order+quartic diagonals	40	0.017	0.17	0.001	0.16	7.2
S_o (sum of squared ab initio derivatives)			188,575,304				

[a] Energy in kcal/mol; distances in Å; angles in degrees; frequency in cm^{-1}.

a diagonal to a full quadratic force field, the deviation of the calculated energies from the ab initio values is reduced from 2.18 to 1.0 kcal/mol, and the deviation with respect to the harmonic spectrum is reduced from 156 to 88 cm^{-1}. The deviations with respect to the bond lengths and angles are also reduced, although these observables are reasonably calculated even at the diagonal level.

This observation with respect to the cross terms is somewhat contrary to the common belief that such terms are important for evaluating the harmonic spectrum but not for the energetics and geometry. At the same time it is clear that to obtain a reasonable fit for the harmonic spectrum (deviations in frequencies significantly lower than 100 cm^{-1}), as well as for the energetics and the equilibrium geometry, both anharmonicities and cross terms should be included (force fields 5 and 6). Thus, the results tend to show that the force field has to be balanced with respect to both types of interactions.

Out-of-Plane Representation, using the ab initio Energy Surface to Probe the Functional Form of Potential Function

The final example we present involves the study of the representation of the energy of out-of-plane distortions [5]. We now turn to the question of how to define the out-of-plane coordinate. Most force fields used in simulating proteins and other biomolecular systems define this coordinate as an improper torsion. This definition is somewhat artificial and the choice of the dihedral angle about

Table 5 *Example of use of the technique to determine the functional form of potential functions. Comparison of alternative representations of the out-of-plane deformation*[a]

Out-of-plane definition	Parameters	Distorted configurations[b]		Equilibrium geometry[b]		
		$\sqrt{S/S_0}$	ΔE, rms	Δr, rms	$\Delta\theta$, rms	$\lvert\Delta\nu\rvert$, avg
Pyramid height	40	0.017	0.17	0.001	0.16	7.2
Improper torsion	40	0.032	0.53	0.001	0.45	13.4
Wilson et al. (symmetrized)	40	0.018	0.21	0.000	0.11	11.8

[a] Force field 6.
[b] Symbols and units as in Table 4.

which the torsion is measured is arbitrary. As a result, the out-of-plane coordinate is ambiguous. Alternative definitions are the height of the pyramid formed by the four atoms (i.e., the distance coordinate discussed in the beginning of this section) or the definition given by Wilson et al. [41] (i.e., the angle between a bond and the plane formed by the other two bonds). The pyramid height definition is independent of the ordering of the atoms and therefore is unambiguous. Table 5 shows results obtained with the fully anharmonic force field (6) by using all three definitions. As demonstrated, the improper torsion definition yields the worst performance and results in energy deviations that are a factor of two to three times larger than those obtained with the other two definitions. First and second derivatives are also not fit as well and will affect the dynamics, as reflected in the larger deviations in harmonic frequencies.

This case is a specific example of the importance of the functional form used in the force field. All three force fields listed in Table 5 are identical, except for the definition of this coordinate (see Fig. 4), and consequently they have the same number of parameters. The differences in performance stem directly from differences in the quality of approximation of the out-of-plane coordinate.

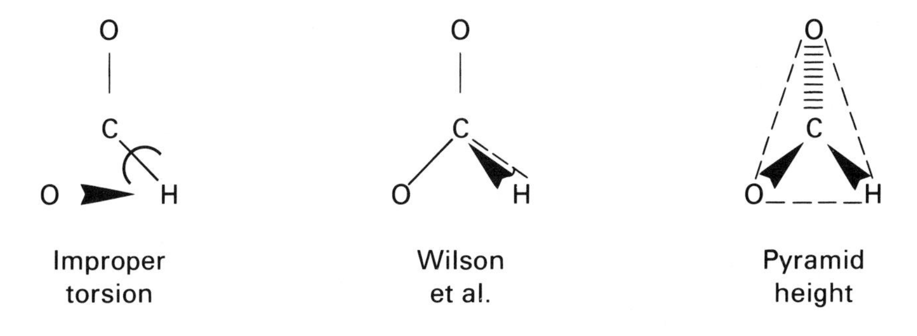

Fig. 4. Definitions of the out-of-plane coordinate.

Thus, this demonstrates how the method proposed in this paper is particularly suitable for resolving questions of the appropriate functional form to represent the energy of a molecule as a function of the coordinates. In other words, the procedure is highly useful for determinations of the 'laws of force' for intramolecular deformations.

Acknowledgements

This work was supported by the Consortium for Research and Development of Potential Energy Functions (including Abbott Laboratories, Battelle Pacific Northwest Laboratory, BIOSYM Technologies, Convex Computer Corporation, Cray Research, E.I. DuPont de Nemours & Co., ETA Systems, IBM, ICI, Merck, Sharp & Dohme Research Laboratories, Monsanto Company, Rohm and Haas Company, Sandoz, Takeda Chemical Industries, and The Upjohn Company).

References

1. Hagler, A.T., In Hruby, V.J. and Meienhofer, J. (Eds.) The Peptides, Vol. 7, Academic Press, New York, NY, 1985, pp. 213-299.
2. Hagler, A.T. and Moult, J., Nature, 272 (1978) 222.
3. Karplus, M. and McCammon, J.A., CRC Crit. Rev. Biochem., 9 (1981) 293.
4. Dauber, P. and Hagler, A.T., Acc. Chem. Res., 13 (1980) 105.
5. Maple, J.R., Dinur, U. and Hagler, A.T., Proc. Natl. Acad. Sci., U.S.A., in press.
6. Hagler, A.T., Huler, E. and Lifson, S., J. Am. Chem. Soc., 96 (1974) 5319.
7. Hagler, A.T. and Lifson, S., Acta Crystallogr., Sect. B, 30 (1974) 1336.
8. Hagler, A.T., Stern, P.S., Lifson, S. and Ariel, S., J. Am. Chem. Soc., 101 (1979) 813.
9. Hagler, A.T., Lifson, S. and Dauber, P., J. Am. Chem. Soc., 101 (1979) 5122.
10. Hagler, A.T., Dauber, P. and Lifson, S., J. Am. Chem. Soc., 101 (1979) 5131.
11. Lifson, S., J. Chim. Phys. Phys.-Chim. Biol., 65 (1968) 40.
12. Momany, F.A., Carruthers, I.M., McGuire, R.F. and Sheraga, H.A., J. Phys. Chem., 78 (1973) 1595.
13. Hagler, A.T. and Lapiccirella, A., Biopolymers, 15 (1976) 1167.
14. Scheraga, H.A., Adv. Phys. Org. Chem., 6 (1968) 103.
15. Ramachandran, G.N. and Sasisekharan, V., Adv. Protein Chem., 23 (1968) 283.
16. Kitaigorodskii, A.I., Molecular Crystals and Molecules, Academic Press, New York, NY, 1973.
17. See for example, International Tables for X-Ray Crystallography, Vol. III, Kynoch, Birmingham, 1962, p. 202.
18. Coulson, C.A., In The Shape and Structure of Molecules (Oxford Chemistry Series), Clarendon Press, Oxford, 1974, p. 37.
19. Harel, M., Ph.D. Dissertation, Weizmann Institute of Science, Rehovot, 1974.
20. Hagler, A.T. and Lifson, S., J. Am. Chem. Soc., 96 (1974) 5327.
21. Scheraga, H.A., Chem. Rev., 71 (1971) 195.
22. Dunning Jr., T.H. and Winter, N.D., J. Chem. Phys., 55 (1971) 3360.
23. Momany, F.A., Carruthers, L.M. and Scheraga, H.A., J. Phys. Chem., 78 (1974) 1995.

24. Williams, D.E. and Starr, T.L., Comput. Chem., 1 (1977) 173.
25. Lifson, S. and Stern, P.S., J. Chem. Phys., 77 (1982) 4542.
26. See 'News', Nature, 322 (1986) 586.
27. Harding, L.B. and Ermler, W.C., J. Comput. Chem., 6 (1985) 13.
28. Pople, J.A., Krishnan, R., Schlegel, H.B. and Binkley, J.S., Int. J. Quant. Chem., 13 (1979) 225.
29. Blaney, J., Weiner, P., Dearing, A., Kollman, P.A., Jorgensen, E., Oatley, S. Burridge, J. and Blake, C., J. Am. Chem. Soc., 104 (1982) 6424.
30. Wipff, G., Dearing, A., Weiner, P., Blaney, J. and Kollman, P.A., J. Am. Chem. Soc., 105 (1983) 997.
31. Brooks, B.R., Bruccoleri, R.E., Olafson, B.D., States, D.J., Swaminathan, S. and Karplus, M., J. Comput. Chem., 4 (1983) 187.
32. Weiner, S.J., Kollman, P.A., Case, D.A., Singh, U.C., Ghio, C., Alagona, G., Profeta Jr., S. and Weiner, P., J. Am. Chem. Soc., 106 (1984) 765.
33. Nilsson, L. and Karplus, M., J. Comput. Chem., 7 (1986) 591.
34. Weiner, S.J., Kollman, P.A., Nguyen, D.T. and Case, D.A., J. Comput. Chem., 7 (1986) 230.
35. Hagler, A.T., Stern, P.S., Sharon, R., Becker, J.M. and Naider, F., J. Am. Chem. Soc., 101 (1979) 6842.
36. Lifson, S. and Warshel, A., J. Chem. Phys., 49 (1968) 5116.
37. Warshel, A. and Lifson, S., J. Chem. Phys., 53 (1970) 582.
38. Ermer, O. and Lifson, S., J. Am. Chem. Soc., 95 (1973) 4121.
39. Kuczera, K. and Czerminski, R., J. Mol. Struct., 105 (1983) 269.
40. Halgren, T.A., Maximally Diagonal Force Constants in Dependent Angle-Bending Coordinates: Implications for the Construction of Model Empirical Force Fields, to be published.
41. Wilson, E.B., Decius, J.C. and Cross, P.C., Molecular Vibrations, Dover, New York, NY, 1980, p. 59.
42. Shimanouchi, T., Tables of Molecular Vibrational Frequencies, Consolidated Vol. I, Natl. Stand. Ref. Data Ser., Natl. Bur. Stand., 1972, pp. 39, 45, 92-93, 126-129, 150.
43. Schachtschneider, J.H. and Snyder, R.G., Spectrochim. Acta, 19 (1963) 117.
44. Snyder, R.G. and Schachtschneider, J.H., Spectrochim. Acta, 21 (1965) 169.
45. Wilmshurst, J.K. and Bernstein, H.J., Can. J. Chem., 35 (1957) 969.

The design of novel proteins using a knowledge-based approach to computer-aided modeling

T.J.P. Hubbard and T.L. Blundell

Laboratory of Molecular Biology, Department of Crystallography, Birkbeck College, University of London, Malet Street, London WC1E 7HX, U.K.

A systematic technique for protein modeling that is applicable to the design of drugs, peptide vaccines and new proteins has been used in the design of a novel protein, CRYSTANOVA. CRYSTANOVA is based on a single domain of the two-domain eye-lens crystallin protein family. It is an attempt at creating a more symmetrical structure than naturally exists, and it incorporates a copper binding activity that is not present in the native protein family. The approach to the design is knowledge-based: it uses the structures of the homologous protein family and sets of rules generated from a wider database of homologous families to rapidly test possible sequences by generating model structures automatically.

Introduction

Advances in recombinant DNA technology, allowing the characterization, production and engineering of new proteins, have provided previously undreamt of possibilities for the construction of novel molecules of industrial, agricultural and clinical interest. However, very few of the almost infinite number of molecules that could be synthesized can have the structural properties to make them useful. A systematic approach is needed to select candidates that have the requisite molecular architecture. For this reason protein crystallography and modeling have become central to rational approaches to: (a) modeling of protein receptors for drugs, herbicides and insecticides; (b) design of protein and peptide vaccines; (c) protein engineering of novel molecules.

The availability of detailed structural information is already providing the information to engineer enzymes with altered specificities [1] and greater stability [2, 3]. The stability of many enzymes outside the cell will need to be increased in order to make their use in industrial and consumer applications more wide spread.

Although X-ray diffraction studies have elucidated the structures of about 400 proteins, this has not led to a full understanding of the rules of protein folding. At present, it is not possible to predict with any degree of certainty the three-dimensional structure adopted by an amino acid sequence, unless there is a strong sequence homology with a protein where the three-dimensional structure is known. Our present knowledge is certainly insufficient to design from scratch completely new proteins that will fold correctly and consequently there has been limited success in this area [4, 5].

A more conservative objective – to construct proteins analogous to those existing in nature – may be feasible with our present knowledge. This exploits rules derived from the three-dimensional structures of related proteins at the level of globular domains, motifs, secondary structure and amino acid side-chain conformations [6-8]. Considerable advantage is gained from the use of a relational database of protein structures [9 and Islam, S., Gardner, S., Sternberg, M.J.E., Thornton, J.M., in preparation]. A similar approach for use in modeling to electron density in protein crystallography allows selection and fitting of substructures from a database of high resolution structures during interactive computer graphics sessions [10].

When designing novel proteins, yet more variables exist. The sequence whose structure is being predicted can be changed at will and there is no certainty that a particular novel sequence will adopt any stable structure at all. Thus, rather than design in a completely ab initio fashion, current projects have tended to be based on finding a novel sequence that will satisfy an existing protein fold. As it is essential to explore as many alternative models as possible and to consider their likely stabilities, a computer program that builds models automatically for many different sequences is invaluable. It greatly reduces the time needed to model an amino acid sequence, so increasing the number that can be investigated and so improving the chance of identifying a good candidate.

The modeling system COMPOSER [6, 7] was being developed at the same time as work was carried out on the CRYSTANOVA design. Thus we were able to compare the advantages of this automatic computer system to the more interactive and intuitive approach that was also used.

The Rationale behind the CRYSTANOVA Design

CRYSTANOVA is based on the structure of the γ-crystallin family of proteins. These crystallin proteins are major structural constituents of the eye lens of vertebrates and form several families [11, 12]. The three-dimensional structures of γ-II crystallin [13] and γ-IV crystallin [14] have been solved and show that each structure is composed of two very similar domains each composed of two similar motifs (see Fig. 1). The structural similarity, sequence identity and correspondence of exons to motifs in the homologous β-crystallin gene family [17] clearly indicate that the protein arose through two rounds of gene duplication.

Each motif consists of four β-strands. When two motifs are brought together, two β-sheets are formed, each comprising three strands from one motif and

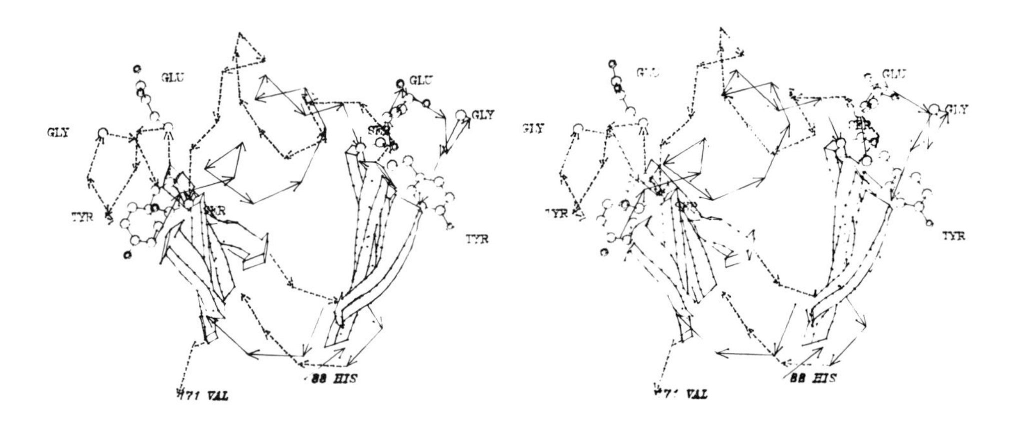

Fig. 1. Stereo diagram of α-carbon atoms of a single crystallin domain with four conserved residue side chains labeled. A dotted line connects α-carbon atoms of second motif. Picture produced by a computer program written by A.M. Lesk and K.D. Hardman [15,16].

one from the other. The most characteristic feature of the crystallin motif is the short loop connecting the first two strands which folds over to bury several conserved residues. Although the structure of the four motifs in a crystallin protein is very highly conserved, only four amino acid residues are conserved throughout the sequences of all γ-crystallin motifs. When the homologous β-crystallins and bacterial S-protein [18] sequences are included in the comparison, only two residues are identical throughout. This extreme flexibility in sequence, coupled with a highly stable structure (witnessed by their stability in the eye lens for a human lifetime) makes crystallin proteins an ideal starting point for a new protein design.

The objective of the design was to mirror evolution after the first of the two hypothetical gene duplications i.e., to design an amino acid sequence for a single domain, where the number of amino acid identities between the two motifs making up that domain was as high as possible (see Fig. 2).

Natural crystallins have no catalytic activity or other feature that can provide a sensitive assay of the integrity of the tertiary structure. Therefore an attempt was made to introduce a symmetrical metal binding site into CRYSTANOVA. Copper was chosen, not because it occurs naturally in such a fold, but because it has excellent spectral properties. The site also has potential use as a marker to select sequences giving improved copper binding by random mutagenesis of the designed sequence. This is possible because copper is toxic to bacteria and so cells containing a protein which bound copper most strongly (whether through mutations that improved binding site geometry or that increased the stability of the overall fold) might be expected to grow best on media containing copper.

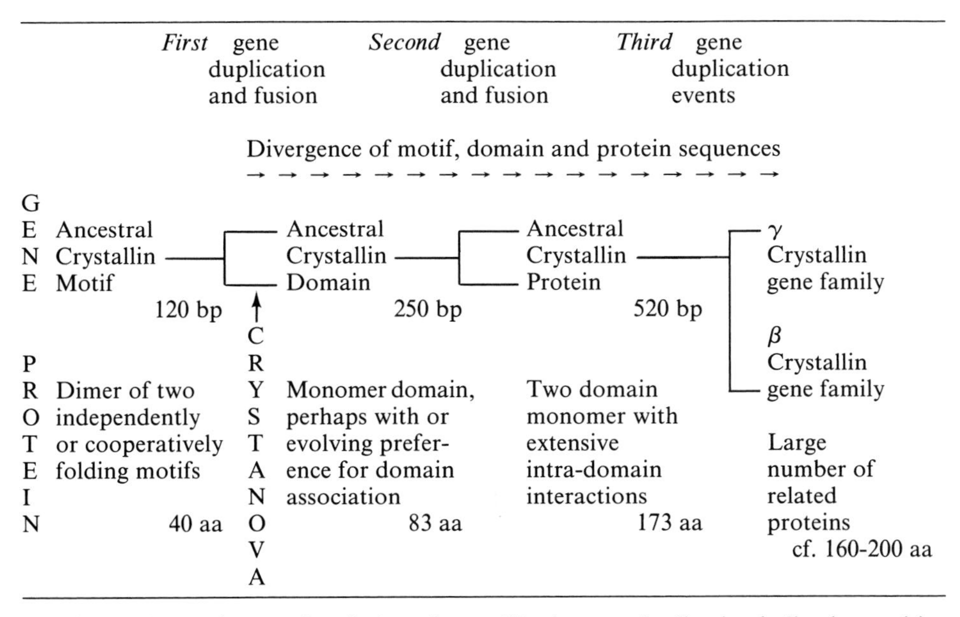

Fig. 2. Phylogenetic tree of evolution of crystallins by gene duplication indicating position of CRYSTANOVA designed sequence on this: qualitative description only.

Steps in the Design of CRYSTANOVA Sequence and Structural Model

Methods used

The coordinate files used for the analysis were taken from the Brookhaven protein database [19] except for the γ-IV crystallin coordinates (kindly provided by Huub Driessen; White, H.E., Driessen, H.P.C., Slingsby, C., Moss, D.S. and Lindley, P.F., in preparation) and the coordinates of the S-protein and β-crystallin proteins, which had been modeled by homology within the Department of Crystallography, Birkbeck College, London. The display and comparison of structures was carried out using the program HYDRA [20] on an Evans & Sutherland PS340. Fragment manipulation was achieved using the display program FRODO (version Tom 4.0 based on Rice 6.0 [10]) on the same system. The accessible contact surface area to solvent was calculated on whole proteins, domains and motifs using the algorithm of Richards and Richmond [21] as previously described [22]. Assessment of packing of side chains was carried out with PACANA (kindly provided by John Moult, CARB, University of Maryland, U.S.A.) and using Connolly surfaces programs [23] supplied with HYDRA. Energy minimization and molecular dynamics were carried out with GROMOS [24].

At various stages the automatic model building program COMPOSER [6, 7] was used to generate a completely symmetrical model by superposition of an existing model onto a 180-degree rotation of itself; this allowed calculation of an average or framework of secondary structural elements. COMPOSER was also used to mutate residues of an existing structure and place them in a 'best guess' initial conformation. This could be done at as many points as were felt necessary at once and was faster and more reproducible than could be achieved by hand, on computer graphics. Combined with other routines, COMPOSER could be used to propagate mutations which were considered satisfactory on one motif onto the symmetrically equivalent position on the other motif. This enabled rapid checking of the effects of mutations on the interactions between the two motifs. Thus it was possible to benefit from the speed of an automatic model building program without losing beneficial modifications to the structure carried out by hand.

Constructing the backbone for the structural model

The sequence of CRYSTANOVA was influenced by knowledge of the conservation of residue types in the alignment of all sequences of crystallin-like domains. However, the main criteria for residue selection was the effect of sequence modifications on the structural model, as investigated using computer graphics. The X-ray structural data for the γ-II and IV crystallins gave coordinates for four crystallin domains*. Since the objective was to design a sequence with the greatest possible sequence homology between motifs, it was assumed that the structure of the domain would be symmetrical. The starting point for the model was therefore the domain with the greatest internal symmetry. The four domain structures were analyzed to find the one with the highest degree of symmetry between the two motifs within the domain. This was done by superposing each domain onto itself rotated through 180° about the pseudo two-fold axis. By measuring the shift between the superposition of motif 2 onto motif 1 and vice versa (using FITTING, [22]) and splitting this shift into components perpendicular to the pseudo two-fold axis (these components do not affect the symmetry between the motifs) and parallel to the axis (this component does affect the symmetry between the two motifs), the degree of symmetry in each domain was calculated. The first domain of each crystallin was found to have a large shift along the pseudo two-fold axis (see Fig. 3). The γ-IV second domain was chosen as the starting point for the model as it was most symmetrical.

*γ-II-A = γ-II-1st domain; γ-IV-A = γ-IV-1st domain;
γ-II-B = γ-II-2nd domain; γ-IV-B = γ-IV-2nd domain.

Designing the binding site and choice of loops

Only structures of two distinct copper binding sites are known at present. These are azurin/plastocyanin [26, 27] and superoxide dismutase [28]. One of the design models produced at the EMBL 'Protein design on Computers' workshop (September, 1986) was COPROP, which was a modified 4-helix bundle modified to contain a copper binding site with minimal amino acid changes [29, 30]. One of the versions of this protein is now being synthesized through expression in *E. coli* (Chris Sander, personal communication). The geometry of the residues providing the copper binding site in plastocyanin was used to search for possible sites in the ROP protein that, after mutation, might act as ligands for copper. β-Carbon atoms in the correct orientation were identified by hand fitting on computer graphics. Various possibilities were investigated and the best chosen on the basis of the least destabilizing effect on the rest of the protein.

Similar techniques were applied to the design of the CRYSTANOVA model. In this case, however, the arrangement of ligands around copper in superoxide dismutase was used as a search model, in an attempt to design a symmetrical binding site, i.e., with two ligands from each motif. Although the azurin/plastocyanin site has four amino acid ligands (His, His, Cys, Met), three are almost planar and much closer (with N-Cu and S-Cu distances of about 2.1 Å) than the fourth (with an S-Cu distance of 3.0 Å). On the other hand, superoxide dismutase has four histidine ligands with an almost tetrahedral (and therefore symmetrical) binding arrangement, although a problem with using histidine ligands is the need to provide a hydrogen bond donor for the NH group not involved in ligand binding.

As can be seen from Fig. 1, the most obvious place to build a symmetrical binding site is between the two loops connecting strands 3 and 4. Anywhere else would either not be symmetrical, or use ligands deeply buried in the structure. Natural copper binding sites are found near the surface of proteins. Figure 3 shows a structural alignment of the sequences of the eight crystallin motifs. The loops connecting strands 3 and 4 are clearly the most variable part of the structure. Since the decision was made to build a symmetrical binding site in CRYSTANOVA between the two loops, the loop selection and binding site design were closely linked. A loop was needed that could:

(1) be included as a connection between β-strand 3 and 4 for both motifs;
(2) provide β-carbon positions that were compatible with binding site residues;
(3) leave no bad contacts or holes at the top of the domain, where the two loops come into contact with each other.

In order to keep the deviation from the natural crystallin structure as small as possible, the initial search was restricted to known crystallin loops, rather

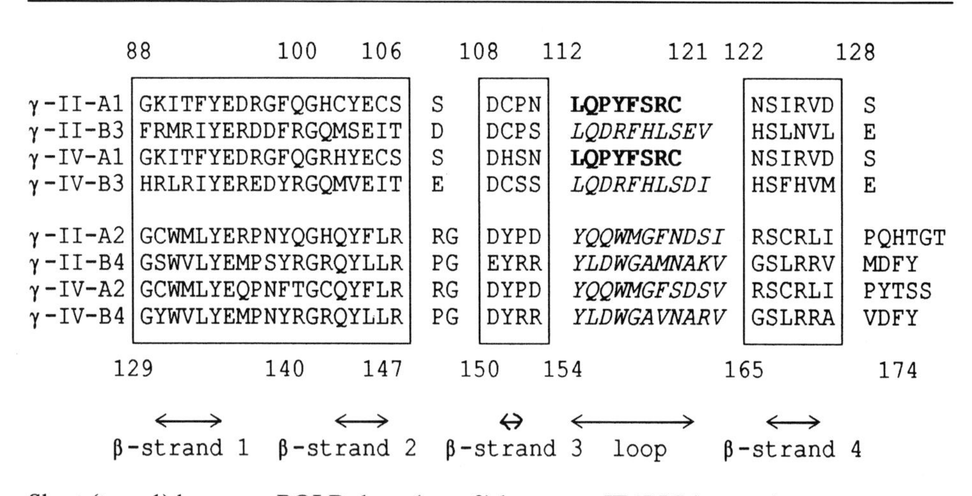

Fig. 3. Structural alignment by motif of known γ-crystallin structures. Topologically equivalent regions are boxed and aligned between motifs. Motifs are labeled 1-4 after domain label (i.e. A1, A2, B1, B2). Numbering 88-128 and 129-174 is for γ-IV-B domain. Strands are as defined for γ-IV-B by DSSP [25].

than from a wider database of structures. Superposition of all eight loops from the four crystallin domains showed that there were two families. A short form (type 1: 9 residues) was found in the first motif of each of the first two domains. A longer form (type 2: 11-12 residues) was found elsewhere. Although there was variation in length and chain tracing, all loops occupied similar regions in space with respect to contacts with the rest of the protein. This suggested that any combination of the two families was possible, i.e., in the first domains a type 1-type 2 interaction is seen; in the second domain a type 2-type 2 interaction is seen: therefore, a type 1-type 1 interaction should also be possible.

Initially it was decided to investigate a model with two type-2 loops since type-2 loops were more common and this combination was found naturally. Unfortunately, it proved very difficult to find β-carbon positions that were satisfactory for a copper binding site. Several designs were made, but each buried the binding site deep inside the protein, which was felt too unrealistic or at least too unpredictable in this case. The use of two type-1 loops produced a much more satisfactory arrangement and was the selection adopted in the final design.

Substitution of side chains

In previous work [22], residues were identified where the solvent accessibilities of side chains were less than 7% of the value for extended side chains. Solvent accessibility calculations were carried out on the whole protein, individual domains and individual motifs of each of the crystallin structures. From consideration of the differences between these results for each residue, groups of residues in different environments were identified. General guidelines governing substitutions for the three groups identified were as follows:

(1) *Solvent inaccessible in all calculations*: These residues are buried in the motif structures themselves: their existence supports the idea that the motif may be a pseudo stable folding unit of the protein and that changes could be made in the interaction between motifs without affecting a large part of the folding pathway of the protein. These residues were not changed.

(2) *Solvent inaccessible in domains but not motifs*: These residues are buried by the interaction between the two motifs and are necessary for the formation of a single domain. Substitutions to other hydrophobic residues to improve symmetry were allowed.

(3) *Solvent inaccessible only in whole protein calculation*: These residues are buried at the domain-domain interface in γ-II and IV. Since CRYSTANOVA was designed to be a single domain protein, these residues were marked for possible substitution to types found at topologically equivalent positions on the other motif not involved in domain-domain interactions.

The mutations that were made to the sequence of the main chain elements of crystallin domains initially selected are shown in Fig. 4.

The first design (CN8) using the loop arrangement described above had a reasonable histidine ligand organization, although two of the distances were rather too long. It was hoped that the arrangement adopted in a real protein of that structure might be more satisfactory and that it would be possible to refine the arrangement using energy minimization and molecular dynamics techniques. A major problem was the identification of hydrogen bond donors for the nonliganding histidine NHs. A perfunctory search of the Brookhaven files identified a few examples of threonine-O-γ in hydrogen bonding distance of a histidine-NH where both were buried. This was felt to be sufficient evidence that the design of such a hydrogen bonding arrangement could be justified.

No further modifications of internal residues were carried out at this stage, other than the substitution of leucine to threonine residues to form hydrogen bonds to the histidine-NH not involved in copper binding, and the substitution of leucines to isoleucine, and vice versa, to increase symmetry between the two motifs and improve interactions between the buried side chains of the β-sheets and the loops.

```
Motif 1

          88          100    106      108   112          121 122     128
γ-IV-B1   HRLRIYEREDYRGQMVEIT   E     DCSS  LQDRFHLSDI   HSFHVM   E
              ↓  ↓     ↓ ↓             ↓↓   ==========   ↓ ↓
CN8       HRLRTYEQEDYRGHMYEIT   E     DIPS  LQPYHSRH     RSTHVM   E
              ↓                        ↓ ↓  ↓   ↓          ↓
CN12      HRLRLYEQEDYRGHMYEIT   E     DLPR  HQPYFSRH     RSTRVM   E
          1       10       19         21    25      32   33        39

         <--->       <--->           <>    <------->       <--->
     β-strand 1  β-strand 2    β-strand 3   loop      β-strand 4

Motif 2

         129         140    147      150   154            165         174
γ-IV-B2   GYWVLYEMPNYRGRQYLLR   PG    DYRR  YLDWGAVNARV   GSLRRA   VDFY
           ↓ ↓↓  ↓     ↓  ↓ ↓           ↓   ===========   ↓ ↓
CN8       GNWRTYEQPNYRGHQYELS   PG    DFRR  LQPYHSRH      RSTRRA
              ↓   ↓↓     ↓  ↓↓   ≡     ↓↓   ↓   ↓
CN12      GNWRLYEQEDYRGHMYEIT   Q     DLPR  HQPYFSRH      RSTRRA
         40        50      58         60    64      71    72   77
                                            ↑       ↑
γ-II-A                                     (LQPYFSRC)
```

*Fig. 4. Sequence changes from γ-IV-B domain to CRYSTANOVA CN8 and CN12. Italic numbering is for γ-IV-B, other is CRYSTANOVA. Loop used for CN8/12 is from **γ-II-A** domain (original sequence shown) and replaces original γ-IV-B loops.*
↓ indicates mutation between topologically equivalent residues.
≡ indicates regions where structures are not quivalent.

The nonsymmetrical but topologically equivalent Leu 3 en Trp 42 buried residues were not changed. The only arrangement considered possible at this stage was to substitute Leu 3 for Trp. This second tryptophan could be accommodated in the model to fill the space created between two of the threonines which had previously been leucines. However, it was felt that such a change was too radical and risky as part of a designed sequence that was going to be built. Also the packing arrangements for the two tryptophans in this model did not reflect those found naturally and were not symmetrical. In any case, a model with two tryptophans may not reflect a previous evolutionary stage: the S-protein [18], which from model building seems to have a higher similarity (at least in size) between buried residues in the two motifs, has alanine or valine at these positions. Such a mutation might be an interesting experiment at a later date, however.

Analysis of the surface residues showed a complex arrangement of possible interactions between the many arginines, glutamates, glutamines, aspartates and histidines on the surface. Many of these interactions were between the two loops and from loops to the rest of the protein. Since the loops were taken from motif 1 of the first domain, and the rest of the protein was from the second domain, careful consideration was given to maintaining interactions seen in both cases. At the same time, interactions between residues within the same loop or β-strand were avoided and the symmetry in sequence between the two motifs was maintained as much as possible.

Tests of the model-built structure

The model was analyzed for bad contacts at 2.5 Å and these were removed by bond rotations by hand. Energy minimizations were used afterwards to improve the overall packing of side chains. An analysis of predicted main chain hydrogen bonds showed that they existed where expected, as in the native domains, but this was unsurprising since such regions of the structure had been little changed. Analysis of packing within the structure found internal holes below the binding site. This was not satisfactory and an attempt was made to use molecular dynamics on the model in a truncated octahedral box of solvent, to look for other side-chain conformations with closer packing arrangements. Whilst the structure of the main part of the protein did not change during this run, the loop arrangement changed significantly and the binding site distances became worse. An analysis of the hydrogen bond lifetimes during the run showed that two of the threonine residues were not interacting significantly with their corresponding histidine-NH.

Although caution is needed in the interpretation of molecular dynamics results, particularly over as short a run as was possible here, it appeared that the designed active site/hydrogen bonds were unlikely to exist in the real protein as initially modeled. However, the movements during molecular dynamics suggested an alternative symmetrical position for two of the histidine ligands. This provided hydrogen bonding to two alternative carbonyl oxygens in the loop itself. This option was modeled to give much more realistic distances for the copper binding ligands and better hydrogen bonds. The return to the original phenylalanines and leucines where there had been histidines and threonines, filled the internal cavity present in the previous design (CN8, Fig. 4). So far there has been insufficient time to carry out molecular dynamics on this version but other analysis of the structure suggested it to be considerably better than before. This sequence (CN12, Fig. 4) was used for the gene design and subsequent cloning.

The final model

Figure 5 shows an α-carbon model of the final structure with active site histidine

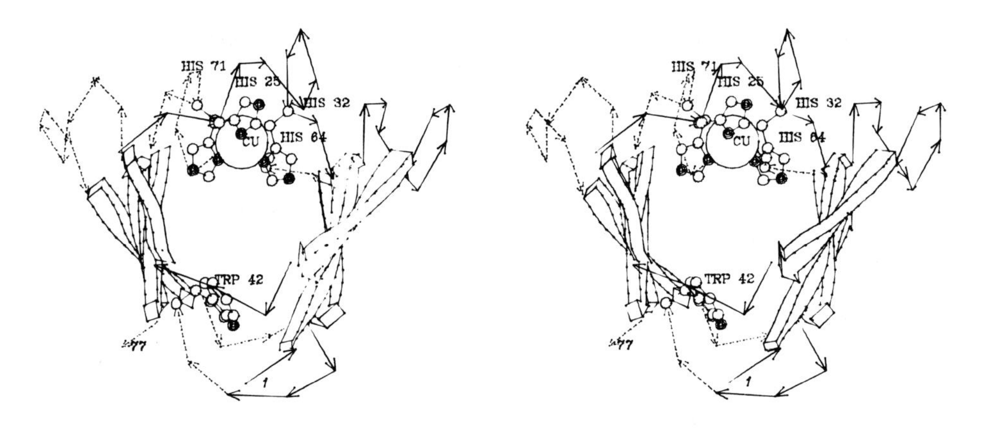

Fig. 5. Stereo diagram of α-carbon atoms of CRYSTANOVA with four histidine copper ligand side chains and nonsymmetrical tryptophan labeled. A dotted line connects α-carbon atoms of second motif. Picture produced by a computer program written by A.M. Lesk and K.D. Hardman [15,16].

side chains shown and the nonsymmetrical tryptophan. Regions of β-sheet conformation are shown as a ribbon model. The motif/motif sequence similarity has increased from 20-30% identities to 84% identities from natural domains to CRYSTANOVA.

Table 1 shows the percentage identities and sequence homology (according to a Dayhoff score) of the topologically equivalent residues in pairs of the domains considered in Fig. 3. It can be seen that the highest similarities are between the 1st domains of γ-II/IV and between the 2nd domains of γ-II/IV as expected, as they are most closely related evolutionarily [Johnson, M.J., Sutcliffe, M.J., Blundell, T.L. (1988), unpublished results]. CRYSTANOVA is most similar to

Table 1 *Homology and identities for model compared with four γ-crystallin domains of known structure (84% residues were topologically equivalent). Topologically equivalent residues were defined by MNYFIT [6,7] and FITTING [Hubbard, unpublished program]*

Homology (according to Dayhoff) \ Identical residues (%)	γ-II-A	γ-II-B	γ-IV-A	γ-IV-B	CRYSTANOVA model
γ-II-A	–	41.5	86.2	41.5	40.0
γ-II-B	107	–	38.5	75.4	46.2
γ-IV-A	131	102	–	41.5	35.4
γ-IV-B	106	129	103	–	58.5
CRYSTANOVA model	102	108	100	111	–

Table 2 *Distribution of amino acids in CN12 and γ-crystallin like domains*

Amino Acid	Protein	%	No.	0% 5% 10%	Amino Acid	%	No.	0% 5% 10% 15% 20%
CYS	CN12	0.0	(0)		TYR	9.5	(8)	
	γ-II-A	7.1	(6)			9.5	(8)	
	γ-II-B	1.2	(1)			8.3	(7)	
	γ-IV-A	6.0	(5)			9.5	(8)	
	γ-IV-B	1.2	(1)			0.7	(9)	
	S-Pro-A	0.0	(0)			3.6	(3)	
	S-Pro-B	1.2	(1)			3.6	(3)	
MET	CN12	3.6	(3)		ASP	4.8	(4)	
	γ-II-A	2.4	(2)			7.1	(6)	
	γ-II-B	6.0	(5)			8.3	(7)	
	γ-IV-A	2.4	(2)			7.1	(6)	
	γ-IV-B	3.6	(3)			8.3	(7)	
	S-Pro-A	1.2	(1)			4.8	(4)	
	S-Pro-B	2.4	(2)			7.1	(6)	
ILE	CN12	2.4	(2)		PRO	4.8	(4)	
	γ-II-A	4.8	(4)			6.0	(5)	
	γ-II-B	2.4	(2)			3.6	(3)	
	γ-IV-A	3.6	(3)			4.8	(4)	
	γ-IV-B	3.6	(3)			2.4	(2)	
	S-Pro-A	8.3	(7)			6.0	(5)	
	S-Pro-B	4.8	(4)			7.1	(6)	
VAL	CN12	2	(1)		GLN	6.0	(5)	
	γ-II-A	1.2	(1)			8.3	(7)	
	γ-II-B	6.0	(5)			3.6	(3)	
	γ-IV-A	2.4	(2)			7.1	(6)	
	γ-IV-B	7.1	(6)			3.6	(3)	
	S-Pro-A	10.7	(9)			6.0	(5)	
	S-Pro-B	6.0	(5)			4.8	(4)	
LEU	CN12	6.0	(5)		SER	4.8	(4)	
	γ-II-A	4.8	(4)			8.3	(7)	
	γ-II-B	10.7	(9)			8.3	(7)	
	γ-IV-A	4.8	(4)			13.1	(11)	
	γ-IV-B	9.5	(8)			6.0	(5)	
	S-Pro-A	7.1	(6)			6.0	(5)	
	S-Pro-B	7.1	(6)			8.3	(7)	
TRP	CN12	1.2	(1)		THR	4.8	(4)	
	γ-II-A	2.4	(2)			3.6	(3)	
	γ-II-B	2.4	(2)			1.2	(1)	
	γ-IV-A	2.4	(2)			3.6	(3)	
	γ-IV-B	2.4	(2)			1.2	(1)	
	S-Pro-A	0.0	(0)			3.6	(3)	
	S-Pro-B	0.0	(0)			6.0	(5)	
PHE	CN12	2.4	(2)		ASN	1.2	(1)	
	γ-II-A	6.0	(5)			4.8	(4)	
	γ-II-B	4.8	(4)			2.4	(2)	
	γ-IV-A	7.1	(6)			3.6	(3)	
	γ-IV-B	3.6	(3)			2.4	(2)	
	S-Pro-A	3.6	(3)			1.9	(10)	
	S-Pro-B	6.0	(5)			9.5	(8)	

Table 2 *continued*

Amino Acid	Protein	%	No.	Amino Acid	%	No.
GLY	CN12	3.6	(3)	GLU	9.5	(8)
	γ-II-A	9.5	(8)		3.6	(3)
	γ-II-B	7.1	(6)		7.1	(6)
	γ-IV-A	8.3	(7)		3.6	(3)
	γ-IV-B	7.1	(6)		7.1	(6)
	S-Pro-A	8.3	(7)		6.0	(5)
	S-Pro-B	7.1	(6)		4.8	(4)
HIS	CN12	8.3	(7)	ARG	16.7	(14)
	γ-II-A	3.6	(3)		9.5	(8)
	γ-II-B	2.4	(2)		14.3	(12)
	γ-IV-A	2.4	(2)		9.5	(8)
	γ-IV-B	4.8	(4)		15.5	(13)
	S-Pro-A	0.0	(0)		2.4	(2)
	S-Pro-B	0.0	(0)		4.8	(4)
ALA	CN12	1.2	(1)	LYS	0.0	(0)
	γ-II-A	0.0	(0)		1.2	(1)
	γ-II-B	2.4	(2)		1.2	(1)
	γ-IV-A	0.0	(0)		1.2	(1)
	γ-IV-B	3.6	(3)		0.0	(0)
	S-Pro-A	9.5	(8)		3.6	(3)
	S-Pro-B	6.0	(5)		3.6	(3)

the γ-IV second domain (on which it was modeled), although the level of sequence similarity to each domain is fairly similar.

Table 2 shows the proportions of each type of amino acid used in CRYSTANOVA compared to the other four domains and the two domains of the S-protein (S-Pro). No serious differences in amino acid usage are found between the 4 γ-crystallin domains and CRYSTANOVA – the largest variation is in the amount of histidine and is due to the inclusion of four extra histidines in the CRYSTANOVA binding site. The S-protein, which is predicted to have a γ-crystallin-type fold, was included to show the differences between it and the γ-crystallin family including CRYSTANOVA. Major differences in amino acid composition found between the S-protein and the other γ-crystallin sequences are: isoleucine/valine/alanine/asparagine (more in S-protein); tyrosine (less in S-protein); arginine (much less in S-protein) and tryptophan/histidine (none in S-protein).

A synthetic gene duplex for CRYSTANOVA has been designed and synthesized as 16 oligonucleotides. This has been cloned into the M13mp18 sequencing vector in *E.coli* and shown to have the correct nucleotide sequence. The coding fragment has been sub-cloned into an expression Vector pRIT2T which allows expression as a fusion with the protein A of *Staphylococcus aureus* under temperature-inducible control [31].

The fusion protein is expressed and has been purified by affinity chromatography using the property of protein A to bind to IgG. Characterization of the protein product is in progress.

References

1. Wilks, H.M., Smith, C.J., Hart, K.W., Barstow, D.A., Atkinson, T., Lee, T.V., Clarke, A.R. and Holbrook, J.J., In Brew et al. (Eds.) Advances in Gene Technology: Protein Engineering and Production (ICSU Short Reports, No. 8), 1988, pp. 154.
2. Creighton, T.E., BioEssays, 8 (1988) 57.
3. Kellis Jr., J.T., Nyberg, K., Sali, D. and Fersht, A.R., Nature, 333 (1988) 784.
4. Rath, V.L. and Fletterick, R.J., Cell, 49 (1987) 583.
5. Van Brunt, J., Bio/Technology, 4 (1986) 277.
6. Sutcliffe, M.J., Haneef, I., Carney, D. and Blundell, T.L., Prot. Eng., 1 (1987) 377.
7. Sutcliffe, M.J., Hayes, F.R.F. and Blundell, T.L., Prot. Eng., 1 (1987) 385.
8. Blundell, T.L., Carney, D., Gardner, S., Hayes, F., Howlin, B., Hubbard, T.J.P., Overington, J., Singh, D.A., Sibanda, B.L. and Sutcliffe, M.J., Eur. J. Biochem., 172 (1988) 513.
9. Blundell, T.L., Sibanda, B.L., Sternberg, M.J.E. and Thornton, J.M., Nature, 326 (1987) 347.
10. Jones, T.A. and Thirup, S., EMBO J., 5 (1986) 819.
11. Slingsby, C., Trends Biochem., 10 (1985) 281.
12. Piatigorsky, J., Cell, 38 (1984) 620.
13. Blundell, T.L., Lindley, P., Miller, L., Moss, D., Slingsby, C., Tickle, I., Turnell, B. and Winstow, G., Nature, 289 (1981) 771.
14. White, H.E., Driessen, H.P.C., Slingsby, C., Moss, D.S. and Lindley, P.F., (1988) in preparation.
15. Lesk, A.M. and Hardman, K.D., Science, 216 (1982) 539.
16. Lesk, A.M. and Hardman, K.D., Methods Enzymol., 115 (1985) 381.
17. Inana, G., Piatigorsky, J., Norman, B., Slingsby C. and Blundell, T., Nature, 322 (1983) 310.
18. Wistow, G., Summers, L. and Blundell, T., Nature, 316 (1985) 771.
19. Bernstein, F.C., Koetzle, T.F., Williams, G.J.B., Meyer, E.F., Brice, M.D., Rodgers, J.R., Kennard, O., Shimanouchi, T. and Tasumi, M., J. Mol. Biol., 141 (1977) 441.
20. Hubbard, R.E., Computer Graphics and Molecular Modelling, In Fletterick, R. and Zoller, M. (Eds.) Current Communications in Molecular Biology, Cold Spring Harbor Laboratory, Cold Spring Harbor, NY, pp. 9-11.
21. Richards, F.M. and Richmond, T.J., Molecular Interactions and Activity in Proteins, CIBA Foundation Symposium 60, New Series, Excerpta Medica, Amsterdam, 1977, pp. 23-47.
22. Hubbard, T.J.P. and Blundell, T.L., Prot. Eng., 1 (1987) 159.
23. Connolly, M.L., Science, 221 (1983) 709.
24. Van Gunsteren, W.F. and Berendsen, H.J.C., Groningen Molecular Simulation (GROMOS) Library Manual, BIOMOS, Nijenborgh 16, Groningen, 1987, pp. 1-229.
25. Kabsch, W. and Sander, C., Biopolymers, 22 (1983) 2577.
26. Adman, E.T. and Jensen, L.H., Isr. J. Chem., 21 (1981) 8.
27. Colman, P.M., Freeman, H.C., Guss, J.M., Murata, M., Norris, V.A., Ramshaw, J.A.M. and Venkatuppa, M.P., Nature, 272 (1978) 319.

28. Richardson, J.S., Thomas, K.A., Rubin, B.H. and Richardson, D.C., Proc. Natl. Acad. Sci. U.S.A., 72 (1975) 1349.
29. Degrado, W.F., Hubbard, T.J.P., Reichelt, J. and Woodward, C., In Sander, C. (Ed.) Protein Design Exercises 1986 (EMBL Biocomputing, Technical Document No. 1), BIOcomputing, EMBL, 6900 Heidelberg.
30. Sander, C., Protein design, EMBL Annual Report, 1986, pp. 132-133.
31. Nilsson, B. Abrahmsen, L. and Mathias, U., EMBO J., 4 (1985) 1075.

Shape-guided generation of conformations for cyclic structures

Paul R. Gerber
Central Research Units, F. Hoffmann La-Roche,
CH-4002 Basel, Switzerland

Introduction and Outline

Finding the low-energy conformations of medium-sized and large ring systems is a classical problem in structural chemical research [1-3]. Even putting aside the question of quality of the underlying model (in our case a molecular mechanics force field) the task of enumerating the total set of potentially interesting structures becomes a formidable one for an increasing number of flexible hinges in the ring. While existing schemes essentially screen a total set of conformations for ring-closing ones, the present new approach tries to characterize the ring conformations geometrically in terms of generic shapes that are expected to exhaust, in a gross fashion, the geometrical possibilities of the ring. The generic shapes are characterized by node numbers for axial and radial deviations from a regular polygon and, in general, violate the chemical requirements of the structure by a fair amount. A subsequent energy minimization procedure takes care of these nonchemical features. In order to allow for the possibility of multiple occurrence of a single generic shape in the final conformations, the generic shapes are taken repeatedly as starting structures and are subjected to a randomization step to avoid duplication.

Generic Ring Shapes and Initial Conformations

The starting point of our description of generic shapes is the regular planar polygon in polar coordinates representation:

$$r_n = r_0,\ \vartheta_n = \frac{\pi}{2},\ \varphi_n = \frac{2\pi}{N} n, \qquad n = 1, \ldots, N \tag{1}$$

In modifying this regular structure to generic shapes, the values φ_n are not varied because such a variation would mainly affect the bond lengths that are quickly readjusted by the force field. The significant features of generic shapes are generated by varying φ_n and r_n. These variations are classified in terms of modes characterized by node numbers k and j [4]:

$$\vartheta_n = \frac{\pi}{2} + \Theta_j \cdot \sin(\frac{2\pi}{N} jn + \delta_j), \qquad 2 \leqslant j \leqslant \frac{N}{2}$$

$$r_n = r_0 + R_k \cdot \sin(\frac{2\pi}{N} kn + \zeta_k), \qquad 2 \leqslant k \leqslant \frac{N}{2} \tag{2}$$

Values of k and j above N/2 are excluded because of the discreteness of the variable n. Furthermore, $j=0$ only moves the centroid and $j=1$ tilts the ring, while $k=0$ corresponds to a change in scale and $k=1$ shifts the polygon.

The remaining j- and k-values lead to characteristic shape distortions of the ring, each having a set of phase angles, δ_j and ζ_k, respectively, associated with it. Since the phases are restricted to the interval $-\pi < \delta_j \leqslant \pi$ and the spacing of values is reasonably on the order of $2\pi j / N$, there are roughly N/j δ-values required to exhaust the relevant node positions.

The ring shapes s are generated by setting up expressions of the form

$$\vartheta_n^s = \frac{\pi}{2} + \sum_j \Theta_j^s \sin(\frac{2\pi}{N} jn + \delta_j^s)$$

$$r_n^s = r_0 + \sum_k R_k^s \sin(\frac{2\pi}{N} kn + \zeta_k^s) \tag{3}$$

with specific values of the amplitudes Θ_j^s, R_k^s and the phases δ_j^s, ζ_k^s.

Since the subsequent minimization procedure removes and modifies most of the fine details that may be put into the start conformation, it is important to extract a set of starting parameters that select just the essential shape-determining features, Numerous investigations led us to settle on keeping only a single mode in ϑ and r for each initial shape, i.e., only a single j- and k-component in Eqs. 2 and 3 have non-zero amplitude with the values

$$\Theta_j = 2/j$$

$$\frac{R_k}{r_0} = 1 - \mathrm{Min}(\frac{\sqrt{2}}{2}, \frac{10}{N}) \tag{4}$$

The phases run through all the N/j and N/k values as indicated above. Thus, for large values of N the number of generic shapes, S, used as start structures, increases as

$$S \approx \sum_{j=2}^{N/2} \frac{N}{j} \sum_{k=2}^{N/2} \frac{N}{k} \approx (N \ln N)^2$$

which compares favourably with an exponential growth. However, this should not distract from the fact that the number of conformations actually grows exponentially with the number of flexible hinges in the asymptotic limit. We expect, however, that the randomization step uncovers an increasing number of conformations that can be grouped under the same generic shape, at least in the range of systems that can reasonably be treated at present.

Actually, the algorithm takes the same generic shape after randomization so many times as starting conformation until a predefined number, f, of consecutive trials has not produced a new final structure.

In order to keep the program general enough for practical use, we allow for simple substituents on the ring. Simple chain-like substituents are restricted to a length of just one atom (methyl, carbonyl, etc.) in order not to increase the number of conformations unduly with rotamers of the side chains. Furthermore, (up to six-membered) rings can be included, thus allowing for proline residues and benzo-derivatives. Cyclic peptides containing amino acids with long side chains have to be modeled with alanine as a substitute.

All substituents are added to the randomized ring with standard geometry as far as possible, while stereochemical requirements are satisfied by adding constraints in a suitable form to the force field in the early stage of the minimization.

Minimization and Evaluation of Structures

We will focus on the set of finally generated conformations mostly from a point of view of completeness and leave aside the detailed energetic aspects that are mostly a matter of the applied force field. Nevertheless, a few remarks regarding the requirements on the force field are in order here. Since the start structures generated from generic ring shapes may substantially violate the small-scale aspects of valence geometry, the energy function needs terms to enforce stereochemical features such as cis- or trans-arrangements of double bonds or chirality at centers. Furthermore, the energy function should be free of singularities or problematic functional behavior which may interrupt the minimization procedure. The minimizer must be able to handle highly deformed start conformations efficiently. It should perform smoothly to avoid escaping from shallow secondary minima. A conjugate gradient method with a restart option [5] has proved very useful for our purpose.

Our investigations utilize a special united-atom force field which is implemented in our in-house modeling system MOLOC [unpublished]. It is based on a central force field approximation with usual bond, valence angle and torsion angle potentials. Nonbonded interactions include (1,4)- and (1,n)-type interactions ($n>4$), containing attractive and repulsive components that both remain finite at zero distance and vanish at given threshold distances. A special pyramidality term takes care of the geometrical and chirality aspects at trigonal and tetragonal centers. While electrostatic interactions of the Coulomb type are currently omitted, hydrogen bonding is accounted for by anisotropic geometric potentials. Finally, π-conjugative effects are allowed for via modified Hückel-type bond order calculations. This force field has proved to be very efficient and reasonably accurate for qualitative modeling purposes.

In order to avoid duplicates in the resulting conformations, every newly generated structure is screened against the set of previously generated ones firstly by comparing energies and, if matches within 0.1 kcal/mol are found, by structural comparison. Structural equality is defined by an rms deviation of less than 10 pm after rigid-body superposition [6].

Example: Cyclic Alkanes

As a simple example, the cyclic alkanes are considered, from cyclooctane to cycloheptadecane. The results are summarized in Table 1. Only conformations with energies less than 10 kcal/mol over the ground state are kept. The number of conformations found grows roughly by a factor of two upon inserting an

Table 1 *Number of final conformations N, number of starting structures I, and CPU-time requirements on a microVAX 3500 t as a function of ring size n of cycloalkanes for conformation-generation runs with the present algorithm using a randomization repeat parameter $f=4$*

n	N	I	t(min)	N(H)	I(H)	t(H)
8	4	227	5	4	229	197
9	4	228	7	8	235	271
10	13	383	16	20	397	611
11	19	422	25			
12	47	713	51			
13	113	851	88			
14	200	1286	163			
15	448	1757	320			
16	813	2860	683			
17	1656	4247	1650			

additional link into the ring, in agreement with vague expectations derived from the three torsional energy minima for each bond. A similar increase was found as a result of enumeration on an ideal diamond lattice [7], although the numbers themselves are far below our findings.

The table also contains the number of starting conformations which show that the efficiency grows with the ring size. For the 17-membered ring the efficiency is already so high, that one wonders whether the set of starting structures with our restricted choice of harmonics might still be sufficient for much larger systems.

Figure 1 shows a distribution of conformations over the energy scale. The peaking is characteristic for these systems and in gross agreement with the simple

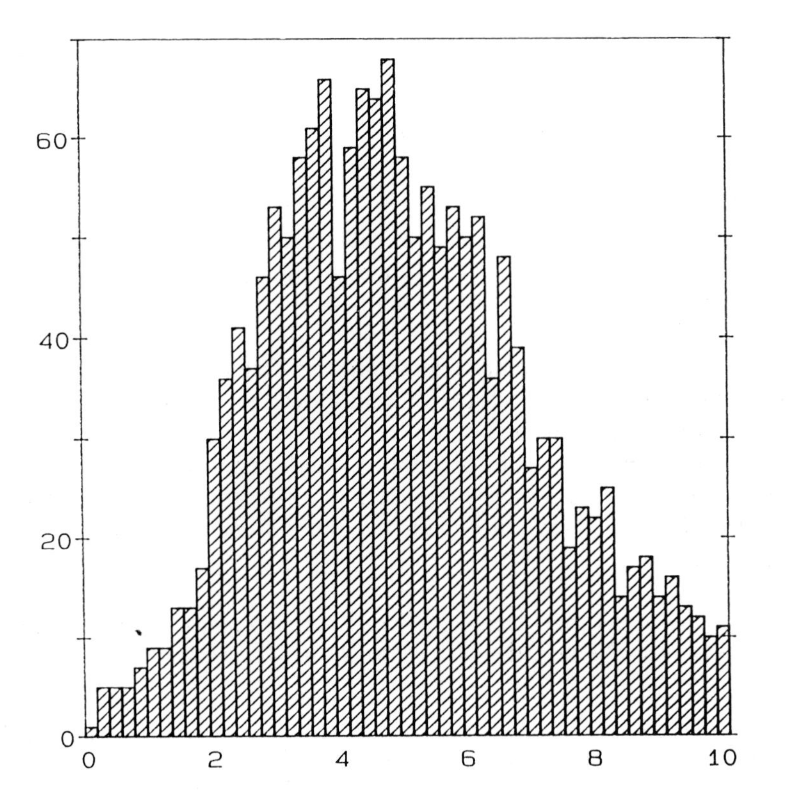

Fig. 1. Histogram of the number of conformations at intervals of 0.2 kcal/mol versus energy (in kcal/mol) for cycloheptadecane. The pattern of low density at the bottom end of the energy scale and of the well-defined maximum is characteristic for these systems, for sizes that admit a fair amount of flexibility.

idea of energy increase of roughly 1 kcal/mol per gauche kink and a combinatorial estimate of the number of arrangements of these kinks on the ring.

For cyclooctane, cyclononane, and cyclodecane a set of structures has been published recently [8]. In order to make a comparison with these results, we ran these systems in the all-atom version. The results are shown in Table 1. We found all the structures of Ref. 8 (with torsional angle agreement within a few degrees) and a few additional ones, which are listed in Table 2. If the resulting structures of the all-atom calculation are minimized in the united-atom force field, they again collapse onto the corresponding smaller sets except cyclodecane where a 7-kcal/mol structure was missed in the united-atom version. One might perhaps ask whether these additional hydrogen-entangled structures are of relevance, considering the sizable zero-point vibrations of the hydrogen atoms.

As can be seen from Table 1, the computational effort increases quite substantially in going from the united-atom approximation to the all-atom version.

Extensions

The algorithm has been extended to treat loop structures protruding from a fixed template. In this case, the boundary condition of having nodes at the anchoring points modifies the rules for possible modes and phase values. The initial phase of minimization now includes, in addition to the steric constraints on the loop, the set of template coordinates as fixed parameters. In the final relaxation stage these may be varied as well, if desired.

An important addition is the possibility to screen the set of resulting structures (loops and rings) for properties of interest. Such programs are at our disposal for various purposes like analysis of secondary structure elements for peptides, comparison with given structures, and so on. Extensions to particular needs can be easily added.

Table 2 *Torsional angles (deg) and energies (in kcal/mol) above the ground state of the structures found in addition to the ones quoted in Ref. 8, when the all-atom version of our force field was used*

n\ Phi	1	2	3	4	5	6	7	8	9	10	Energy
9	−42	−33	−34	120	−15	−79	73	−84	137		13.7
9	61	−44	−54	−1	92	−1	−54	−44	61		17.9
10	94	−128	122	−128	94	−62	103	−170	103	−62	6.1
10	−63	−63	146	−25	−49	−25	146	−63	−63	127	8.4
10	119	−117	85	−45	−176	176	45	−85	117	−119	16.2

Acknowledgements

The present work has profited greatly from cooperation of the author with Klaus Müller and Klaus Gubernator who have put the algorithm to use in practical examples and have helped improve it by their positive criticism.

References

1. Smith, G.M., Program 510 of the Quantum Chemistry Program Exchange, Department of Chemistry, Indiana University, Bloomington, IN.
2. Dygert, M., Gō, N. and Scheraga, H.A., Macromolecules, 8 (1975) 750.
3. Dolata, D.P., Leach, A.R. and Prout, K., J. Comput.-Aided Mol. Design, 1 (1987) 73.
4. Kilpatrick, J.E., Pitzer, K.S. and Spitzer, R., J. Am. Chem. Soc., 69 (1947) 2483.
5. Powell, M.J.D., Mathemat. Progr. 12 (1977) 241.
6. See for example Kabsch, W., Acta Crystallogr., Sect. A32 (1982) 922.
7. Uiterwijk, J.W.H.M., Harkema, S. and van der Waal, B.W., J. Chem. Soc., Perkin Trans., 2 (1983) 1843.
8. Lipton, M. and Still, W.C., J. Comput. Chem., 9 (1988) 343.

Molecular dynamics refinement of the X-ray structure of thermitase complexed with eglin-c

Piet Gros, Masao Fujinaga, Bauke W. Dijkstra, Kor H. Kalk and Wim G.J. Hol
Biomolecular Study Centre (BIOS), Department of Chemistry, University of Groningen, Nijenborgh 16, 9747 AG Groningen, The Netherlands

Protein crystal structure refinement aims at minimizing the differences between the observed structure factor amplitudes, F_{obs}, and the calculated amplitudes, F_{calc}, by shifting the positional and thermal parameters of the N atoms contributing to F_{calc} according to the well-known structure factor equation:

$$F_{\underline{h}} = \sum_j f_j \, e^{-B_j(\sin\theta/\lambda)^2} \, e^{-2\pi i \underline{h}\cdot\underline{x}_j}$$

Here f_j is the scattering factor of atom j, B_j the temperature factor of atom j, $\underline{x}_j$ the position of atom j in fractional coordinates, $\underline{h}$ the indices of the reflection, θ the Bragg angle of reflection j, and λ the wavelength of the radiations used.

One of the first procedures to refine a protein crystal structure was the 'real space refinement' method developed by Diamond [1], which aimed at minimizing $\Sigma(\rho_{obs} - \rho_{calc})^2$ where ρ_{obs} is the best current electron density map, ρ_{calc} the electron density corresponding to the atomic positions of the model and the summation is over all grid points near the atomic positions. In a cyclic process, developed in particular by Deisenhofer and Steigemann [2], the accuracy of protein structures could be greatly improved by this procedure. The method was time consuming, however, but did have a quite good convergence radius when the starting phases for calculating the initial ρ_{obs} were reasonably good. If the initial map was poor, this procedure was very difficult to use.

Since the real space refinement procedure is very CPU-time consuming, Agarwal developed what is called the 'fast Fourier refinement' procedure in which many approximations were made to reduce computational costs, and FFT algorithms for calculating structure factors and derivatives [3]. The function minimized was $\Sigma(F_{obs} - F_{calc})^2$ where the summation is over all observed reflections. Due to a low observation:parameter ratio in virtually all protein structures so far, the coordinates obtained by Agarwal's procedure need to be 'regularized', i.e., given standard geometry in a separate step [4]. The fast Fourier refinement program has not become as widespread as others but the underlying principles have had a great impact on the development of refinement procedures in general.

The most popular refinement procedure used so far in protein crystallography is the 'restrained refinement' method of Hendrickson and Konnert [5]. Here the function to be minimized is $\Sigma(F_{obs} - F_{calc})^2 + \Sigma$ geometry terms $+ \Sigma$ 'other terms'. Weighting factors of these terms with respect to each other, and of the various components of each summation, provided great flexibility to the user. The geometry terms keep bond distances and angles restrained to certain 'ideal' values, keep planar groups planar, etc. 'Other terms' is a collection of restraints such as those relating to temperature factors, 'observed' phases and non-crystallographic symmetry. In spite of all its sophistication and elegance the radius of convergence of this procedure remained limited, almost always below 1 Å and often considerably less. Many computer graphics sessions were therefore needed to rebuild the atomic model in electron density maps.

A procedure quite analogous to the restrained refinement procedure, but incorporating FFT routines as well as more sophisticated energy terms, was that of Jack and Levitt [6]. But, much like the restrained refinement procedure, the convergence radius is below 1 Å and many time-consuming graphics sessions are needed to obtain a well-refined structure.

Recently, Brünger et al. [7] reported on the development of a molecular dynamics refinement procedure where a much larger volume of reciprocal space is searched for solutions because of the kinetic energy of the system. The forces acting on the atoms in the molecular dynamics run are derived (i) from the 'conventional' force field consisting of bond distances, angles, dihedrals plus Lennard-Jones and electrostatic interactions, and (ii) from an X-ray term, namely the derivatives of the sum $\Sigma(F_{obs} - F_{calc})^2$. The kinetic energy of the system allows the exploration of a large number of conformations, quite different from the least squares procedures employed in previous methods where coordinates can only shift along lines of steepest descent. Brünger et al. [7] report encouraging results as do Fujinaga et al. [8] who implemented a molecular dynamics refinement module into the GROMOS molecular dynamics package [9]. Here, several peptide units of a 'wire model phospholipase A_2' structure were flipped in the correct orientation, a change which involves overcoming an energy barrier and is very rarely accomplished by conventional protein structure refinement programs.

Encouraged by these results, we decided to test the power of the molecular dynamics X-ray structure refinement in the case of thermitase, complexed with its inhibitor, eglin-c. Thermitase is a thermostable member of the subtilisin family of serine proteases [10, 11]. Its amino acid sequence is 46% identical to that of subtilisin Carlsberg, having seven additional N-terminal residues, one C-terminal addition, three internal deletion sites (involving seven residues in total), one two-residue insertion and two one-residue insertions. Eglin-c is a 70-residue protein inhibitor, generously provided by Dr. Schnebli, Ciba-Geigy, Basel.

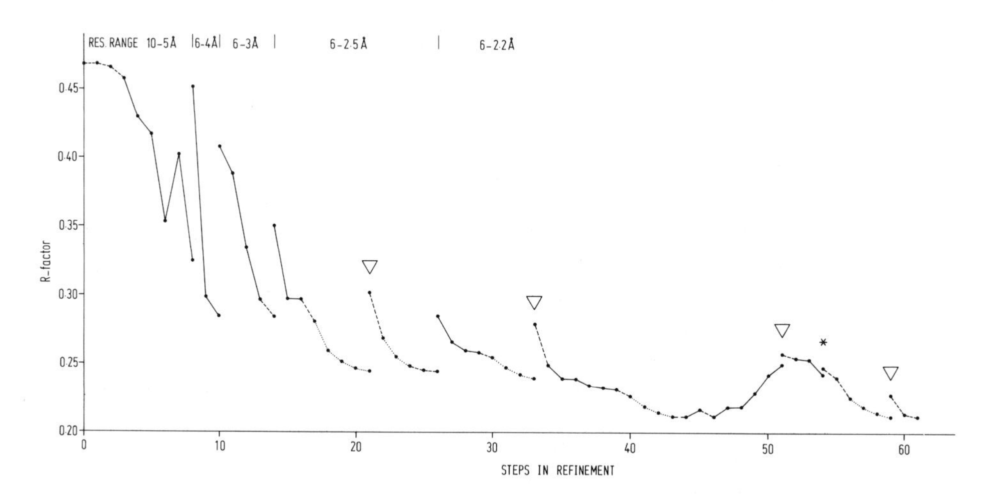

Fig. 1. Course of the refinement as indicated by the crystallographic R-factor.

Crystals of the complex have space group $P2_12_12_1$ containing one thermitase-eglin complex with a molecular weight of ~34 kDa, in the asymmetric unit. A 2.2 Å resolution data set was collected from one crystal on a FAST television area detector diffractometer [12, 13] followed by applying the molecular replacement procedure for an initial structure determination using the Carlsberg-eglin-c complex [14] as a starting model. After a few initial fast Fourier refinement steps an R-factor of 47% was obtained, where

$$R = 100 \times \frac{\Sigma \mid F_{obs} - F_{calc} \mid}{\Sigma F_{obs}} \%.$$

As shown in Fig. 1, this R-factor dropped fully automatically to 25% for all data between 6 and 2.5 Å resolution. After four model building sessions, of which only the first involved some major rebuilding, an R-factor of 21% was obtained for the data between 6 and 2.2 Å. Perhaps the most remarkable feature was that, in several cases, atoms had moved from about 10 Å away towards their correct position. This is a substantial improvement compared with previous refinement methods. Another striking observation was the improvement of the density even in regions where the atoms had not moved automatically to their correct position. This allowed much faster rebuilding in a much better map than was available before applying the molecular dynamics X-ray refinement procedure.

The method worked very well indeed in the case of the thermitase-eglin-c complex where the starting model had serious errors. It is most likely that this new procedure will be used widely in refinements of other biomacromolecular crystal structures.

References

1. Diamond, R., Acta Crystallogr., Sect. A,27 (1971) 435.
2. Deisenhofer, J. and Steigemann, W., Acta Crystallogr., Sect. B,31 (1975) 238.
3. Agarwal, R.C., Acta Crystallogr., Sect. A,34 (1978) 791.
4. Dodson, E.J., Isaacs, N.W. and Rollett, J.S., Acta Crystallogr., Sect. A,32 (1976) 311.
5. Hendrickson, W.A. and Konnert, J.H., In Srinivasan, R. (Ed.) Biomolecular Structure, Function, Conformation and Evolution, Vol. 1 (Int. Symposium Madras, January 1978), Pergamon Press, Oxford, pp. 43-57.
6. Jack, A. and Levitt, M., Acta Crystallogr., Sect. A,34 (1978) 931.
7. Brünger, A.T., Kuriyan, J. and Karplus, M., Science, 235 (1987) 458.
8. Fujinaga, M., Gros, P. and van Gunsteren, W.F., J. Appl. Crystallogr., (1988) in press.
9. Van Gunsteren, W.F. and Berendsen, H.J.C., BIOMOS, Biomolecular Software, Laboratory of Physical Chemistry, University of Groningen, Groningen.
10. Meloun, B., Baudys, M., Kostka, K., Hausdorf, G., Frömmel, C. and Höhne, W.E., FEBS Lett., 183 (1985) 195.
11. Frömmel, C. and Höhne, W.E., Biochim. Biophys. Acta, 670 (1981) 25.
12. Arndt, U.W., Nucl. Instrum. Meth., 201 (1982) 21.
13. Renetseder, R., Dijkstra, B.W., Kalk, K.H., Verpoorte, J. and Drenth, J., Acta Crystallogr., Sect. B,42 (1986) 602.
14. Bode, W., Papamokos, E. and Musil, D., Eur. J. Biochem., 166 (1987) 673.

Biomolecular structures from NMR: Computational aspects

R. Kaptein, R. Boelens and J.A.C. Rullmann
Department of Chemistry, University of Utrecht, Padualaan 8, 3584 CH Utrecht, The Netherlands

Introduction

Over the last decade, NMR spectroscopy has emerged as an important tool for structure determination of biomolecules in solution. Apart from advances in magnet technology, this is to a large extent due to the spectacular improvements in the speed and data storage capacity of modern computers. This has allowed the development of powerful two-dimensional NMR methods [1] which, in turn, made it possible to solve the resonance assignment problem [2], a bottleneck in any detailed NMR study of biomolecules.

The role of the computer in NMR is twofold. First, computers control the experiment - the complex sequence of radiofrequency pulses, delays, phase switching, data acquisition and storage of large amounts of data in the form of free induction decays (FIDs). Secondly, computers are necessary for data processing, which occurs more and more off-line, and for the actual computation of structures based on NMR data. It is to be expected that in all these areas the demands on computer power will only become larger. Especially with the recent development of three-dimensional NMR [3, 4] there is a need for even faster computers interfaced with graphic display systems.

NMR structures are based primarily on a set of short proton-proton distances obtained from the nuclear Overhauser effect (NOE) [5]. The origin of the NOE is dipolar cross-relaxation between protons. Because of the weak proton magnetic moment and the r^{-6} distance dependence of the effect, NOEs can only be measured between protons at relatively short distances (<5 Å). Using suitable calibration procedures, the NOEs can be translated into constraints on proton-proton distances. Several computational methods exist now for structure determination using these distance constraints, such as, for instance, distance-geometry [6, 7] and restrained molecular dynamics [8-10]. As the NOE is a spin-relaxation phenomenon it depends upon the dynamic behavior of the molecule in solution. Therefore, the structure and the dynamics of a biomolecule as seen by NMR are intimately connected. Indeed, NMR is unique in providing information on the dynamics of molecules, as has been realized for a long time.

Protocol for protein structure determination from NMR

---

1. Assign ^{1}H resonances.

2. Determine proton-proton distance constraints and dihedral angle constraints from NOEs and J-couplings, respectively.

3. Calculate family of structures using geometric constraints only (experimental constraints plus covalent structure) using, for instance, distance geometry (DG) and distance bounds driven dynamics (DDD).

4. Refine these structures using geometric constraints and potential energy functions, for instance, with restrained energy minimization (REM) and restrained molecular dynamics (RMD).

Scheme 1

In this article we shall review the methods available for structure determination of biomolecules by NMR. As an example, the determination of the solution conformation of a DNA-binding protein, the *lac* repressor headpiece, will be discussed. Finally, NMR is quite powerful in studies on interactions of biomolecules with all kinds of ligands and the interaction of *lac* headpiece with operator-DNA exemplifies this.

Three-dimensional Structures from NMR

The protocol that we have found useful for structure determination of proteins based on NMR data is shown in Scheme 1. The first two steps, consisting of ^{1}H resonance assignment and determination of distance and dihedral angle constraints, are common to all procedures that have been proposed. Steps 3 and 4 are suitable to address questions such as how unique are the structures obtained, how well do they satisfy the experimentally derived constraints, and how reasonable are they from the point of view of energetics.

In the following, we shall discuss the various steps of Scheme 1 in some detail.

^{1}H resonance assignments

A necessary requirement for the structural analysis of a protein is the assignment of the great majority of its proton resonances. For small proteins ($MW < 10\,000$) that do not aggregate at millimolar concentrations, this can be accomplished using a combination of various 2D NMR experiments. The procedure for the so-called sequential assignment of protein ^{1}H NMR spectra is extensively described by Wüthrich [2]. Briefly, two main classes of 2D experiments are used. In the first, off-diagonal cross-peaks arise only between protons connected through

J-coupling networks, with COSY (correlated spectroscopy) as the prime example. Another very useful experiment in this category is TOCSY (total correlation spectroscopy) [11] or 2D HOHAHA (homonuclear Hartmann-Hahn) [12]. In these, spectra patterns of cross-peaks can be traced between pairs of J-coupled protons as in COSY or between several protons within an amino acid chain as in HOHAHA.

In the second class of 2D NMR experiments, cross-peaks connect protons that are spatially in close proximity (distance <5 Å). The 2D NOE or NOESY experiment [13] and its rotating frame counterpart ROESY [14, 15] fall in this category. A 2D NOE spectrum is recorded in a three-pulse experiment [13].

$$90° - t_1 - 90° - t_m - 90° - t_2 \text{ (acq.)}$$

In this sequence, 90° stands for a 90° radiofrequency pulse; t_1 and t_2 are the variable times which after double Fourier transformation yield the ω_1 and ω_2 frequency domains of a 2D spectrum; t_m is a fixed time, which allows exchange of magnetization between nuclei. The origin of the NOE effect is dipolar cross-relaxation, which depends on fluctuations in the orientation and length of the vectors connecting pairs of nuclei. In a rigid molecule these vectors have fixed lengths and reorient by the tumbling of the molecule as a whole. In that case, cross-relaxation rates are proportional to r^{-6} and therefore have a very strong distance dependence. As an example, a 2D NOE spectrum of *lac* repressor headpiece in D_2O is shown in Fig. 1A with a blow-up of the region of the aromatic resonances in Fig. 1B. All off-diagonal intensity in this spectrum corresponds to short distances between nonexchangeable protons.

The sequential assignment usually starts with a search for cross-peak patterns belonging to the spin-systems of types of amino acids. These are then connected through cross-peaks in a 2D NOE spectrum between neighboring amino acids in the polypeptide chain. Useful short distances involving backbone protons that are manifested in 2D NOE spectra are those between C_α, C_β and amide protons of one residue and the amide proton of the next residue ($d_{\alpha N}$, $d_{\beta N}$ and d_{NN}, respectively). Often the sequential assignment procedure is redundant so that many internal checks are possible. This makes the assignment unambiguous. However, it does not lead to stereospecific assignments for diastereotopically related protons such as those of methylene groups or methyl groups of valine and leucine. Sometimes stereospecific assignments of these protons are possible using a combination of vicinal J-coupling and NOE information [16, 17].

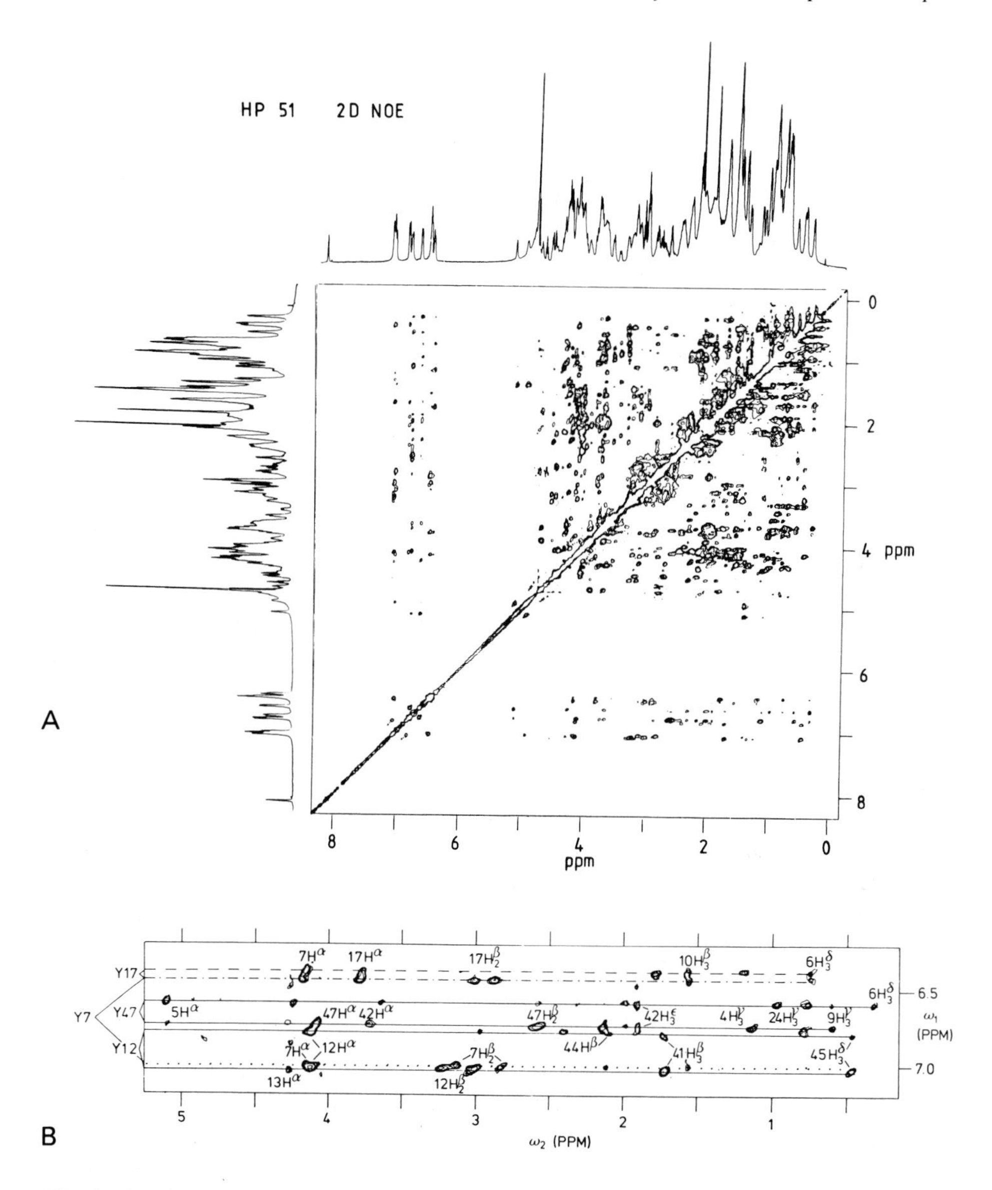

Fig. 1. (A) 2D NOE spectrum of headpiece 51 represented as a contour plot. The spectrum was recorded at 500 MHz of a 5 mM solution of headpiece in D_2O. The mixing time t_m was 100 ms. Off-diagonal cross-peaks indicate short distances (<4 Å) between nonexchangeable protons. (B) Expanded region of the NOE cross-peaks involving the aromatic protons.

Distance and dihedral angle constraints

Proton-proton distance constraints are most conveniently derived from cross-peak intensities in 2D NOE spectra. The initial build-up rate of these cross-peaks in spectra taken at short mixing times is proportional to the cross-relaxation rate σ_{ij} between protons i and j. Therefore, these cross-relaxation rates can be measured either from a single 2D NOE spectrum taken with a sufficiently short mixing time or, more accurately, from a build-up series recorded with various mixing times. For a rigid isotropically tumbling molecule, σ_{ij} is simply related to the distance d_{ij} and the correlation time τ_c:

$$\sigma_{ij} \propto \tau_c \, d_{ij}^{-6} \qquad (1)$$

Therefore, using a known calibration distance d_{cal}, the proton-proton distances follow from the relation

$$d_{ij} = d_{cal} \, (\sigma_{cal} / \sigma_{ij})^{1/6} \qquad (2)$$

In practice, Eqs. 1 and 2 are only approximately valid. There are two main problems associated with accurate determination of proton-proton distances. The first is that of indirect magnetization transfer or 'spin-diffusion'. In reality the NOE cross-peaks are the result of multispin relaxation and only in the limit of extremely short mixing times (where the signal-to-noise ratio is poor) the two-spin approximation of Eq. 1 is valid. The effect of spin-diffusion can be calculated and procedures based on a full relaxation matrix treatment are being developed to solve this problem [18, 19]. Secondly, proteins are not rigid bodies and intramolecular mobility leads to nonlinear averaging of distances and to different effective correlation times for different interproton vectors in the molecule. Internal motions occur over a wide range of time-scales and only the fast fluctuations (up to a few hundred picoseconds) can presently be simulated by molecular dynamics calculations. The slower motions are difficult to handle and therefore constitute the most serious source of error in distance determination from NOEs. For this reason, the approach that is usually taken is that of translating the NOE information into distance ranges (e.g., 2-3 Å, 2-4 Å, 2-5 Å for strong, medium and weak NOEs, respectively) rather than attempting to obtain precise distances. Alternatively, as was done in the case of *lac* headpiece [9], a single cut-off distance is used at 3.5 Å (corresponding to a range of 2-3.5 Å).

In case stereospecific assignments of methylene and methyl groups are not known, the distances involving these protons have to refer to pseudoatoms and a corresponding correction term has to be added allowing for the maximum possible error. For instance, for a methylene group the pseudoatom position is defined at the geometric mean of the CH_2 proton positions and a correction of 0.9 Å is added to the distance constraints involving these protons.

The absence of NOEs between assigned protons also contains useful information as it means that the distance between the protons is larger than ca. 4-5 Å. This non-NOE information has been used in the work on the *lac* headpiece structure [20]. Great care should be taken, however, in interpreting the absence of NOEs, since it may also result from local mobility.

In favorable cases, dihedral angle constraints can be obtained from three-bond J-couplings. These can be obtained from the fine structure of cross-peaks in COSY-like spectra recorded with high digital resolution [17]. An important example is the three-bond coupling $^3J_{HN\alpha}$ between amide and C_α protons, which gives a measure of the backbone torsion angle ϕ. For helical regions $^3J_{HN\alpha}$ is small (ca. 4 Hz), while for extended-chain conformations, such as in β-sheets, it takes large values (9-10 Hz). Usually the large J-couplings (8-10 Hz) are the most useful source of information, because J-couplings smaller than the line-width (typically 5 Hz or larger) cannot be reliably measured due to cancellation effects in antiphase multiplets [21]. Furthermore, the interpretation of the larger coupling constants in terms of dihedral angles is less ambiguous.

Structure calculations based on geometric constraints (distance geometry, distance bounds driven dynamics)

The metric matrix distance geometry (DG) algorithm [6, 22-24] was known well before protein structure determination by NMR became possible. Thus far, it is the only method that does not rely on some starting conformation and is therefore free from operator bias. The DG procedure amounts to the following. First, upper and lower bound matrices U and L are set up for all atom-atom distances of the molecule. Some of the elements u_{ij} and l_{ij} follow from standard bond lengths and bond angles of the covalent structure, and from experimentally found distance ranges from NOEs and J-coupling constants. A bound smoothening procedure using triangle inequalities is applied to these constraints. Then, a distance matrix D is set up with distances chosen randomly between upper and lowerr bounds, $l_{ij} \leq d_{ij} \leq u_{ij}$. The so-called 'embedding' algorithm then finds a 3D structure corresponding to the distances d_{ij}. This structure must be optimized using an error function consisting of chirality constraints (chiral centers sometimes come out the wrong way) and a distance-constraint error function.

This forces the amino acid side chains to adopt the correct chirality and the distances to satisfy the upper and lower bound criteria, although usually the DG structures still contain a number of violations of the distance bounds. By repeating the DG calculations several times the random step of choosing the distance matrix D between upper and lower bounds allows different structures to be obtained and therefore it can be judged how uniquely the structure is determined by the constraints.

Another method, also termed distance geometry but using a quite different mathematical procedure, was suggested by Braun and Gō [7]. Here, the protein conformation is calculated by minimizing a distance constraint error function. Special features of the method are that dihedral angles are used as independent variables rather than Cartesian coordinates and that it uses a variable target function, first satisfying local constraints (between amino acids nearby in the polypeptide chain), while at a later stage long-range constraints are included. Usually one starts with various initial conformations obtained by taking random values for the dihedral angles. A comparison between a metric matrix distance-geometry algorithm DISGEO [25] and the variable target function algorithm DISMAN [7] has shown that the efficiency and convergence properties of both methods are rather similar [26].

Although the distance-geometry method does not need starting structures and is therefore not subject to operator bias, this does not mean that it samples the allowed conformational space (consistent with the bounds) in a truly random fashion. In fact, it was noted by Havel and Wüthrich [24] in model calculations on bovine pancreatic trypsin inhibitor that the DG structures tend to be somewhat more expanded than the crystal structure from which the constraints had been derived. We have also noticed this effect in our work on *lac* headpiece [27]. In regions that are relatively unconstrained, the DG algorithm tends to produce extended conformations for the backbone and side chains (see below). It was further noted that in restrained molecular dynamics refinement a larger variation of conformations was found in regions with few NOEs than with DG, in spite of the fact that the structures now have to satisfy criteria of low potential energy as well. This led to the idea of improving on the sampling properties of the DG procedure by adding a simplified MD calculation with only geometric constraints as the driving potential. In this so-called distance bounds driven dynamics (DDD) algorithm [28], Newton's equations of motion

$$m_i \ddot{\mathbf{r}}_i = \mathbf{F}_i \tag{3}$$

are solved with the forces given by

$$\mathbf{F}_i = -\frac{\partial V}{\partial \mathbf{r}_i} \tag{4}$$

where the potential function now does not contain any energy terms but is taken proportional to the DG error function

$$V = K_{dc}\left[\sum_{d_{ij}>u_{ij}} (d_{ij}^2 - u_{ij}^2)^2 + \sum_{l_{ij}>d_{ij}} (l_{ij}^2 - d_{ij}^2)^2\right] \tag{5}$$

The time-step for integration of Newton's equations (3) should be chosen in accordance with the magnitude of the 'force constant' K_{dc}(taken somewhat arbitrarily as 10 000 kJ mol^{-1} nm^{-4}. Of course, the entirely nonphysical nature of the potential V means that the calculations cannot be interpreted as a simulation of a physical process. A DDD run using a DG structure as the starting conformation has the effect of 'shaking up' the DG structure and thereby improving the sampling of conformation space, which is especially important in regions with few constraints.

Structure refinement including energy terms (restrained energy minimization and molecular dynamics)

The quality of the protein structures based on geometric constraints can be improved by taking energy considerations into account. For instance, in DG structures amino acid side chains often adopt eclipsed conformations, while in alkyl chains the energy of the staggered conformation is 10 - 15 kJ mol^{-1} lower. Also, hydrogen bonds and salt bridges may not be formed unless they are specifically introduced as constraints. In restrained molecular dynamics (RMD), refinement structures are optimized simultaneously with respect to a potential energy function and to a set of experimental constraints. Of course the success of this method now depends to a large extent on the quality of the force field used. It is therefore important to realize the limitations and approximate nature of this force field, especially when the calculations do not include solvent molecules.

In RMD calculations, Eqs. 3 and 4 are integrated with the potential energy function given by

$$V = V_{bond} + V_{angle} + V_{dihedr} + V_{vdW} + V_{coulomb} + V_{dc} \tag{6}$$

The first two terms tend to keep bond lengths and bond angles at their equilibrium values. V_{dihedr} is a sinusoidal potential describing rotations about bonds; for V_{vdW} (the van der Waals interaction) usually a Lennard-Jones potential is taken, and $V_{coulomb}$ describes the electrostatic interactions. The extra distance constraint term V_{dc} distinguishes RMD from more conventional MD simulations. It has the effect of pulling protons within the distance d_{ij}^{o} (or the distance range) in accordance with the NOE observations. Usually a harmonic pseudopotential is chosen

$$V_{dc} = \frac{1}{2} K_{dc} (d_{ij} - d^{o}_{ij})^2 \quad (7)$$

Other forms of V_{dc} are a half-harmonic potential [9] and pseudopotentials that contain a linear part at long distances and repulsive terms for non-NOEs [27]. Similarly, sinusoidal terms describing dihedral angle constraints from J-coupling constants can be included [27].

A restrained dynamics calculation is usually preceded by restrained energy minimization (REM) using steepest descent or conjugate gradient methods to bring the energy down to an acceptable level. REM using the same potential energy function (Eq. 6) usually changes the conformation only slightly and cannot take it out of local minima. By contrast, RMD is able to overcome barriers of the order of kT because of the kinetic energy in the system and therefore has a much larger radius of convergence. RMD works as an efficient minimizer, since excess potential energy, converted to kinetic energy, is drained off by coupling the system to a thermal bath of constant temperature. It has been suggested [29] that the RMD procedure could be used to obtain folded protein structures starting, for instance, from a fully extended polypeptide chain. Although apparently this was successful for model calculations on crambin [29], it is our experience that this procedure does not work in general and is certainly not cost-effective in terms of the best use of computer time. In our view, RMD should be considered as a structure refinement tool using DG or DDD structures as starting conformations.

A Protein Structure: *lac* Repressor Headpiece

The DNA binding properties of the *E.coli lac* repressor reside in the N-terminal domain or headpiece. By proteolytic digestion, headpieces of 51, 56, or 59 amino acid residues can be prepared depending on the enzyme used. These headpieces retain their folded structure and base pair recognition properties when separated from the tetrameric repressor core. In spite of many attempts, no crystals have been prepared of sufficient quality for X-ray crystallography, so that all structural information on the protein has come from NMR. Because the *lac* repressor-operator system has been extensively studied by biochemical and genetic methods, it lends itself very well for a biophysical study. In the early 80s, we started an NMR investigation with the aim of elucidating the structure of *lac* headpiece and its mode of binding to operator-DNA.

The first step in the structure determination of *lac* headpiece consisting of the 51 N-terminal amino acid residues of *lac* repressor was the sequential assignment of its ^{1}H resonances according to the procedure described by Wüthrich [2]. For all amide and C_α-protons, except those of Ile 48 (due to an overlap problem), assignments could be made [30, 31]. Most of the side-chain protons have also been assigned, although for some of the long side chains of Lys,

Fig. 2. Amino acid sequence of the lac *repressor headpiece and summary of the data used for locating the secondary structure. Strong and weak NOEs are distinguished by the thickness of the lines. Filled squares indicate amino acids for which slow H-D exchange was found for the backbone protons.*

Arg and Gln the assignments do not extend beyond the C_β-protons [30, 31]. By making use of a combination of J-coupling ($^3J_{\alpha\beta}$) and NOEs, the pro-chiral methyl groups of valines 9, 20, 23 and 38 could be stereospecifically assigned [16].

Inspection of assigned cross-peaks in the 2D NOE spectra of *lac* headpiece showed that it is a typical α-helical protein with no β-sheets. In three regions of the protein, relatively strong NOEs were found that correspond to short distances d_{NN}, $d_{\alpha N}(i, i+3)$ and $d_{\alpha N}(i, i+4)$ prevailing in α-helices [32]. The three α-helices are found in the regions 6-13, 17-25, and 34-45 of the polypeptide. Consistent with this is the observation of slow H-D exchange for the amide protons in the helical hydrogen bonds [33]. The short- and medium-range NOE and exchange data, on which the secondary structure is based, is summarized in Fig. 2. It should be noted that with the combination of NOE and exchange data the secondary structure of proteins can be established quite reliably.

The next step was the tertiary structure determination. This was done on the basis of 169 NOEs observed in 2D NOE spectra taken at a relatively short mixing time of 50 ms. The distribution of these NOEs is shown in Fig. 3. It can be seen that there are several NOEs connecting the N-terminal and C-terminal regions and, furthermore, that all three α-helices are interconnected. As a distance calibration we used the NOE intensities of a series of cross-peaks corresponding to the distances $d_{\alpha N}$ and $d_{\alpha N}(i, i+3)$ of α-helical regions, which are both approximately 3.5 Å. Thus, the NOEs were essentially converted to distance ranges of 1.8-3.5 Å (the lower limit being the sum of the van der Waals radii). To these ranges, pseudoatom corrections were applied for groups that show dynamic averaging effects (methyl groups and tyrosine rings), and in cases where the stereospecific assignments of diastereotopically related protons were not known. To this data set, 17 H-bond constraints were added for those slowly exchanging amide protons for which the H-bond acceptor was known, i.e., for the α-helical regions. The initial structure determination was carried out with the restrained molecular dynamics method, using a model-built conformation as the starting structure [9, 34]. A relatively low value for the force constant

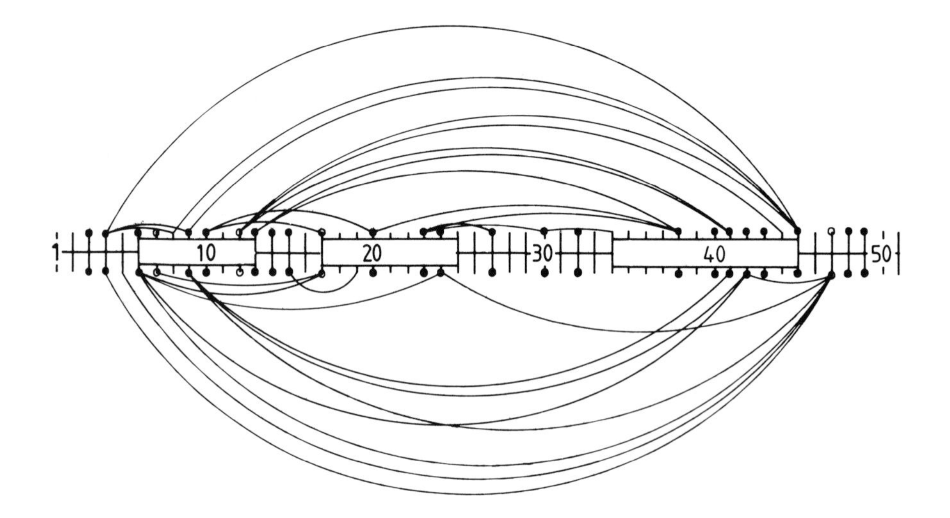

Fig. 3. Survey of the long range NOE connectivities of lac *repressor headpiece. The three helices are indicated by boxes. Nonpolar side chains are marked with filled circles and tyrosyl residues by open circles.*

K_{dc} of 250 kJ mol^{-1} nm^{-2} was used, allowing for excess distances ($d_{ij} - d^{o}_{ij}$) of ca. 1 Å before an energy penalty is felt at the level of kT. However, larger values for K_{dc} (up to 4000 kJ mol^{-1} nm^{-2}) were sometimes found to speed up convergence. At a later stage the absence of NOEs between assigned protons was also taken into account [20]. Thus, repulsive pseudopotential terms for a carefully selected set of 9529 non-NOEs were included. Although the number of non-NOEs may seem large, the information content is in fact rather low, since most non-NOEs are trivial in a structure that is already approximately correct. After an RMD run of 60 ps (without solvent), the resulting structure satisfied the experimental constraints very well and, at the same time, had a low energy (the energy dropped from +4074 kJ mol^{-1} for the starting structure to −3091 kJ mol^{-1} after RMD followed by REM). The remaining violations of the constraints were not larger than 0.5 Å, while the sum of all violations was 5.8 Å [20]. Figure 4 shows a stereopicture taken from the RMD run. The helix-turn-helix region consisting of the helices I and II of the headpiece can be clearly seen with the third helix packing against these forming a hydrophobic

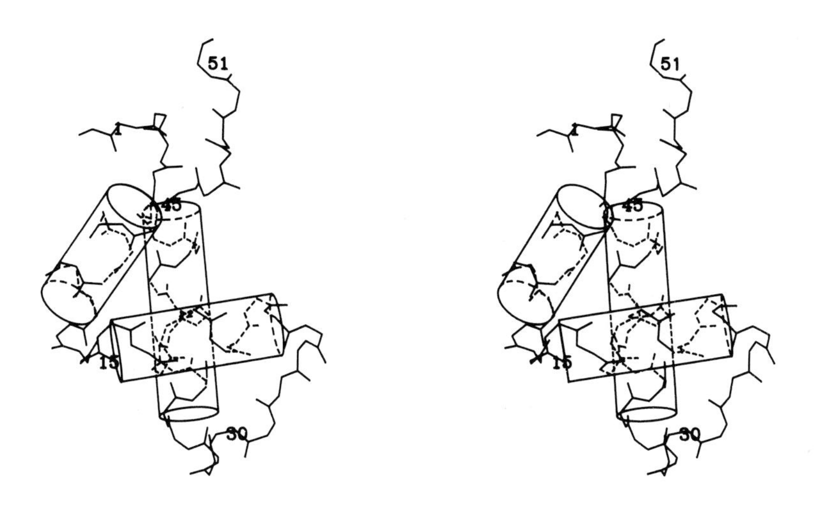

Fig. 4. Stereodiagram of the backbone conformation of headpiece 51. Cylinders represent the three α-helices. The structure was determined from a set of 169 proton-proton distance constraints from NOEs using a restrained molecular dynamics procedure [9].

core. The RMD run also indicated that the three-helical core of the protein is rather rigid, whereas the N-terminal and C-terminal regions and also the loop between helix II and III showed higher mobility.

Next, the question of uniqueness was addressed or, in other words, what is the range of conformations consistent with the constraints. A series of DG calculations was performed using the same set of distance constraints [27]. An overlay of 10 DG structures is shown in Fig. 5A. The variation among these structures can be expressed as an average root mean square (rms) difference of the C_α-atom coordinates, which was 1.4 Å for all C_α's and 1.1 Å for those of residues 4-47. It can be noticed in Fig. 5A that the N-terminal and C-terminal peptide regions have a preference for extended backbone conformations in spite of the fact that there are no long-range NOEs in these regions that would fix the conformation with respect to the helical core. This is clearly an artifact of the metric matrix DG procedure. Starting with the DG structures, DDD runs were carried out with 1000 integration steps each (formally corresponding to 2 ps). Figure 5B clearly shows that the N-terminal and C-terminal peptide fragments adopt a much wider range of conformations, while the structures satisfied the constraints equally well. The average rms differences of the DDD structures were 3.0 Å for all C_α-atoms and 2.0 Å for the C_α's of residues 4-47. Thus, the DDD procedure in combination with DG greatly improves the sampling properties compared to DG alone.

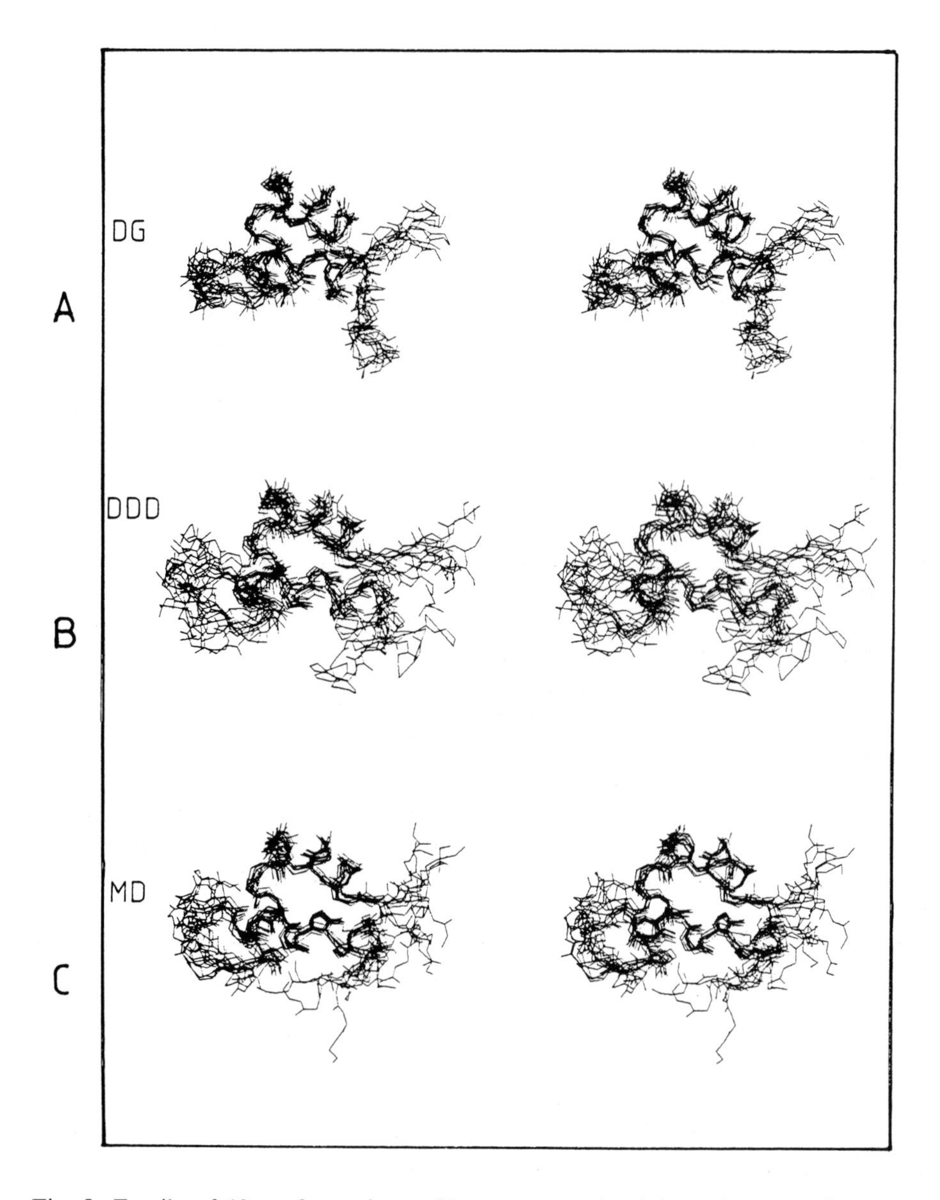

Fig. 5. Family of 10 conformations of lac *repressor headpiece obtained after distance geometry (A), distance bounds driven dynamics (B), and restrained molecular dynamics refinement (C). Stereodiagrams of the backbone atoms are shown.*

Finally, RMD refinement of these structures resulted in the family of structures shown in Fig. 5C. Some convergence can be noticed in helical regions; the rms difference for the C_α-atoms of residues 4-47 now becomes 1.7 Å (for the C_α's of the helices this value is 0.8 Å showing that the helical core of the protein is particularly well determined). Calculated for all C_α-atoms the rms difference is still 3.0 Å which is mainly caused by the large spread in conformations of the N-terminal and C-terminal peptides.

Protein DNA Interaction: *lac* Headpiece-Operator Complex

NMR studies

The *lac* headpiece (HP) structure formed the basis for further studies aimed at determining the way in which *lac* repressor recognizes its operator. *Lac* operator of *E.coli* is defined genetically as the control region in the *lac* operon, where operator constitutive mutants occur. The region protected by *lac* repressor is 20-25 bp (base pairs) long with a pseudodyad axis going through GC 11 [35].

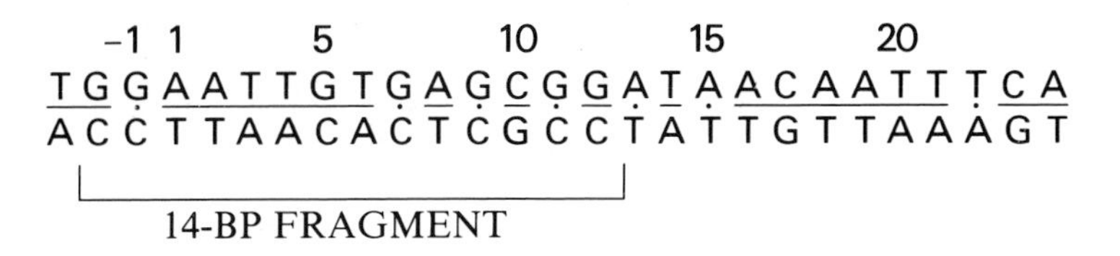

Based on the hypothesis that *lac* repressor binds to the operator with two headpieces, thereby preserving the pseudo C_2 symmetry in the repressor-operator complex, we have chosen a 14-bp half-operator fragment for studies with single headpieces. The total molecular weight of the headpiece - 14-bp operator complex is ca. 14 000, approaching in complexity the limit of what can be presently studied by 2D NMR at the level of individual resonance assignments for large numbers of residues. The first 2D NMR spectra of the complex were already reported in 1983 [36]. However, the detailed analysis took several years [37, 38] and is still not finished.

Figure 6 shows one of the most readily accessible regions of a 2D NOE spectrum of the complex of HP56 with the 14-bp operator. It contains a window, where only intra-DNA cross-peaks occur (H6/H8-H1′ and cytosine H5-H6). These cross-peaks provided a start for the assignment of the DNA resonances, as is shown in Fig. 6 for one strand by the lines connecting intra- and internucleotide cross-peaks. In this way assignments were obtained for all nonexchangeable protons of the DNA in the complex except for some of the H5′ and H5″ protons. The general pattern of intra-DNA NOEs is still that of a B-DNA type conformation. Also, most of the ^{1}H resonance positions show small shifts upon complex

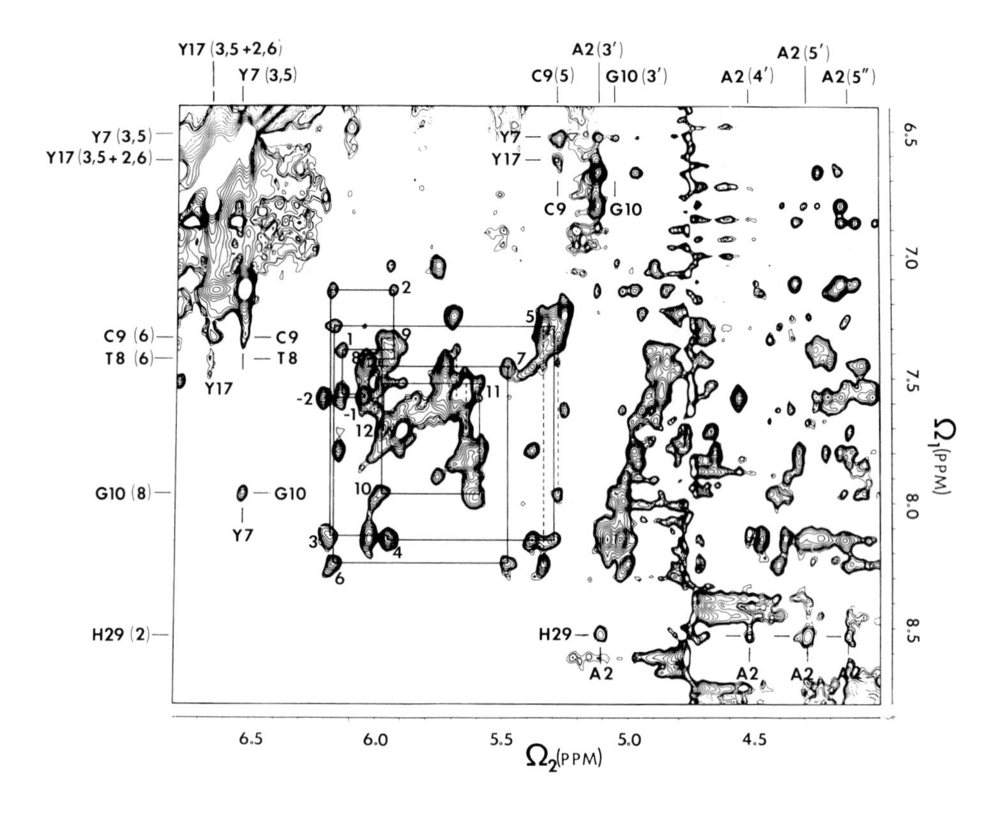

Fig. 6. Part of the 500 MHz 2D NOE spectrum of the HP56 - 14 bp operator complex. Sequential NOEs of one strand of the operator are indicated by the connecting lines. The protein-DNA NOE cross-peaks present in this part of the spectrum are also indicated.

formation with a maximum of 0.36 ppm for the H6 proton of T3. These results are also consistent with the idea that small adjustments of the DNA conformation occur. These conformational changes, however, cannot yet be specified, but are likely to involve bending or unwinding or a combination of both.

Similarly, a large number of ^{1}H assignments has been made for the protein part of the HP 56 - 14-bp operator complex. For this, a combination of 2D NOE spectra and homonuclear Hartman-Hahn (HOHAHA) spectra was used. The protons of many internal residues such as Leu 6, Val 9, Leu 45 and Tyr 47 have characteristic chemical shifts (distinct from random coil), which change very little upon complex formation. Furthermore, ca. 80% of the long-range NOEs could be identified for headpiece in the complex. This shows that the basic three-helical structure of headpiece is conserved when it binds to the operator. Shifts only occur for residues in the DNA binding site, which may be due to the presence of the DNA (for His 29 mainly due to an increase in its pK_a) or to a repositioning of the side chains as probably occurs for Tyr 7 and Tyr 17.

Table 1 *NOEs between* lac *repressor headpiece 56 and a 14-bp* lac *operator fragment*

Protein		DNA	
Unambiguous [a]			
Tyr 7	H3,5	G10	H8
Tyr 7	H3,5	G10	H1′
Tyr 7	H3,5	G10	H3′
Tyr 7	H3,5	C9	H5
Tyr 7	H3,5	C9	H6
Leu 6	CδH3	C9	H5
Tyr 17	H3,5 + H2,6	C9	H5
Tyr 17	H3,5 + H2,6	C9	H6
Tyr 17	H3,5 + H2,6	T8	H6
His 29	H2	A2	H8
His 29	H2	T3	CH3
Probable [a]			
Thr 5	CγH3	G10	H8
Thr 5	CγH3	G10	H3′
Leu 6	CδH3	C9	H5
Leu 6	CδH3	C9	H6
Leu 6	CδH3	C9	H3′
Leu 6	CδH3	T8	H6
Tyr 17	H3,5 + H2,6	T8	CH3
Ser 21	CαH	T8	CH3
Ser 21	CβH	T8	CH3
His 29	H2	A2	H3′
His 29	H2	A2	H4′
His 29	H2	A2	H5′ [b]
His 29	H2	A2	H5″ [b]

[a] The unambiguous NOEs were assigned at unique resonance frequencies, while the probable NOEs were from resonances which could overlap with resonances of other protons (see text for further discussion).
[b] H5′ and H5″ protons were only pair-wise assigned.

Using the assignments of the 14-bp operator and of HP56 in the complex, it was possible to detect NOEs between protein and DNA. Some of these can be seen as cross-peaks in the 2D NOE spectrum of Fig. 6. For instance, extending the horizontal line at 7.96 ppm (of the H8 proton of G10) to low field, one finds a cross-peak at 6.53 ppm, which can only belong to the 3,5 protons of Tyr 7. The analysis so far has yielded 24 inter-protein-DNA NOEs, which are listed in Table 1. A distinction is made between NOEs that are unambiguous because they involve protons with unique resonance positions and those that are probable. The latter ones occur in crowded regions where overlap of resonances may occur. They were assigned on the basis of a pattern-recognition procedure, which involves the following reasoning. Suppose a cross-section of a headpiece proton shows NOEs to a set of other protons of the same amino acid residue. Then, if a cross-section of a DNA proton shows cross-peaks at the same frequencies and at least one of these can be uniquely assigned to a headpiece proton, we consider the assignment of the other cross-peaks in the set also extremely likely. The case of His 29 may serve as an example. In Fig. 6, a horizontal line at the C2 proton frequency of His 29 (8.52 ppm) shows a set of four cross-peaks in the DNA ribose region which are in a crowded region of the spectrum. Now, a similar set of cross-peaks is observed at the line of H8 of adenine 2 (A2) and, moreover, a very weak cross-peak is observed at the crossing of this line and that of His 29, which have both unique resonance positions (not shown). Hence, the weak NOE between H8 of A2 and the His 29 C2 proton is listed in the upper part of Table 1 although it undoubtedly is the result of spin diffusion. The other NOEs of His 29 with the ribose protons of A2 are stronger and represent shorter distances, but occur in a region of the spectrum with much overlap and are therefore indicated as probable.

A structural model for the headpiece-operator complex

Since we have shown that neither headpiece nor operator undergoes large conformational changes upon binding, it seemed reasonable to attempt building a model of the complex based on the structures of the constituents and the NOEs of Table 1. It should be realized that at this stage only a low-resolution model of the complex can be obtained, because the 2D NOE spectra were run under conditions of limited spin diffusion and the distance constraints derived from them could be up to ca. 6 Å. The DNA was kept in the standard B-DNA conformation. For headpiece, the backbone conformation as found by Zuiderweg et al. [34] was also kept rigid, but some of the side chains in the DNA binding site were allowed to change their conformation. It was then possible to obtain a model that satisfies all NOE constraints [37, 38]. Energy minimization of this model (J. de Vlieg and R. Kaptein, unpublished results) was carried out to ensure that it has reasonable nonbonded interactions. A schematic picture of this model is shown in Fig. 7. The fact that all NOE constraints can be accounted for is significant, because it means that the model must represent the major specific complex. Other, nonspecific complexes that are undoubtedly formed at the high concentration (5 mM) of the NMR experiments apparently do not lead to inconsistent NOEs either because they do not involve short proton-proton distances or because each of them does not have a long enough lifetime to allow buildup of NOEs. A 1:1 mixture of two different complexes would have led to inconsistent NOEs and can also be excluded.

The most surprising feature of the model is that the orientation of the second or 'recognition' helix in the major groove of DNA with respect to the dyad axis at GC11 is opposite to what is found in all other models of repressor-operator interaction, either from direct X-ray observation as for 434 repressor [40] or from models built for CAP and λ and cro repressors [41]. Indeed, it is also opposite to orientations predicted for *lac* repressor on the basis of the analogy of models for CAP [42] and cro repressor [43]. In these models, the first helix would be away from the dyad axis, while in the complex shown in Fig. 7 it is close to it. Our model accounts for the phosphate ethylation interference experiments (indicated in Fig. 7) and also for a functional contact between Gln 18 and GC 7 as found by Ebright [44] from a genetic 'loss-of-contact' study involving mutants of both *lac* repressor and operator.

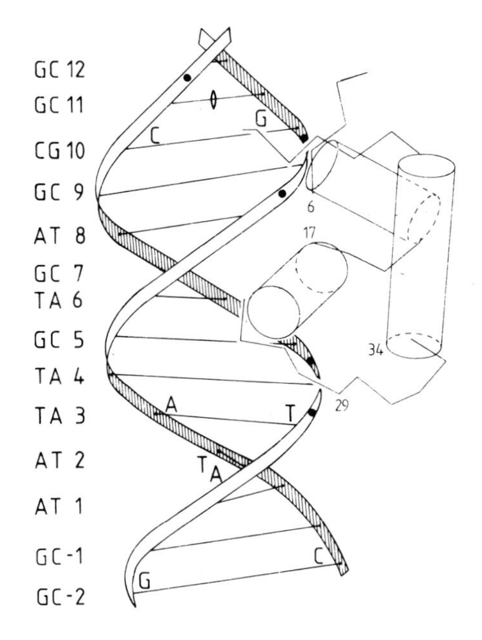

Fig. 7. Model of the headpiece-operator complex. The DNA is in the standard B-conformation, while the backbone conformation of headpiece is taken from Zuiderweg et al. [31]. The black dots indicate the phosphates where ethylation interferes with lac *repressor binding (W. Gilbert and A. Maxam cited in Ref. 39).*

In a recent study of a complex with a 22-bp symmetric *lac* operator, which binds two headpieces simultaneously, the same reversed orientation was found (Lamerichs, Boelens and Kaptein, unpublished results). Although this still does not prove that the whole *lac* repressor binds operator in the same way, there is some recent evidence from genetic experiments that this is in fact the case. Müller-Hill (personal communication) found that a repressor mutant with Arg 22 replaced by Asn has an altered binding specificity, now showing affinity to *lac* operators with base pair GC5 replaced by a TA. This implies a functional contact between Arg 22 and GC5 in the native repressor that would be in accordance with the reversed headpiece orientation.

NMR-based docking of lac *headpiece on* lac *operator*

Finally, the headpiece - 14-bp operator complex was modeled in a more unbiased way than could be done by model building on a graphics display system. For this we used a combination of docking with the ellipsoid algorithm and refinement by restrained molecular dynamics [45]. The ellipsoid algorithm [46] is a special kind of minimization procedure which avoids local minima by making large, discontinuous steps. The variational space consists of 6 docking variables and possibly one or more dihedral angles. Each step is taken within an ellipsoid - defined in this variational space - of gradually decreasing size, in such a way that for convex problems the solution is always contained within the ellipsoid. The optimization ends when the ellipsoid volume or the target function falls below a predefined limit. The method has been shown to be robust in nonlinear problems, such as occur in molecular geometry optimizations subject to distance constraints. One advantage of the ellipsoid algorithm is the possibility to handle only one constraint per step, i.e., steps are executed on the basis of the gradient of a single violated constraint, leading to a considerable saving of time.

Since it had been found that no major conformational changes occur for headpiece or DNA upon binding, the operator was kept in the standard B-DNA conformation and headpiece in the conformation determined previously [35]. Only the amino acid residues 4 - 8 and 16 - 30, which may contact the DNA, were allowed to change their side-chain torsion angles in some of the docking runs.

Distance constraints between protein and DNA were derived from observed NOE cross-peaks (cf. Table 1). Since long mixing times were used, contributions from spin-diffusion are present. Most upper-bound constraints between protein and DNA were therefore set to 6 Å; for some stronger NOEs the corresponding constraint was set to 4 Å. In the case of non-stereospecifically assigned protons, and for situations where dynamic averaging occurs (rotating methyl groups and flipping aromatic rings), the constraint referred to a pseudoatom and the bound was increased accordingly [27].

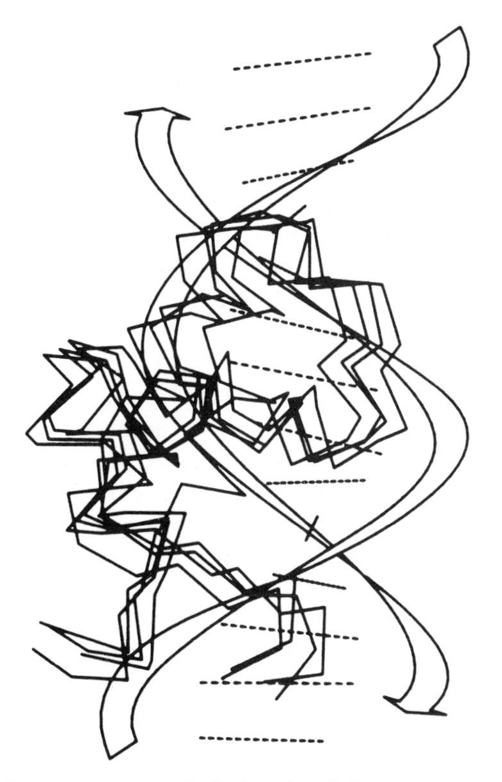

Fig. 8. Overlay of four structures of the lac *headpiece-operator complex obtained by ellipsoid docking and subsequent refinement by restrained energy minimization and molecular dynamics. The DNA was kept in an energy-minimized B-DNA conformation. The rms differences of the C_α-atom positions of the headpiece are within 3.5 Å.*

In the handbuilt model after 5 ps RMD refinement there were still some distance constraints violated (sum of the violations 3.6 Å). By contrast, after ellipsoid docking starting from essentially random orientations, all of the distance constraints could be satisfied. In these structures the headpiece had moved towards the dyad axis by about 1 bp (rms differences of C_α-atom positions were ca. 5 Å with respect to the handbuilt model).

Similar results were obtained with ellipsoid docking in which side-chain dihedral angles were allowed to change. The total number of variables was 66, resulting in a very slow decrease of the size of the ellipsoid. Convergence to total violations of ca. 2 Å in 2 out of 3 runs could be reached by first making 1000 steps starting from a large ellipsoid. After EM, the final structures had small NOE violations (sum $<$ 1 Å) and no violations of the van der Waals distances. Figure 8 shows four structures after alignment of the DNA. The C_α-atom rms differences were up to 3.5 Å, which is a measure of the precision of the structure so far obtained.

While already several hydrogen bonds between protein and DNA were seen in the MD trajectories, some that are critical for base pair recognition such as Gln 18-GC7 and Arg 22-GC5 were not yet formed in the short MD runs employed. In order to see whether these are nevertheless compatible with our model, a numer of MD runs were performed with extra constraints between the side-chain amide of Gln 18 and GC7 and the guanidylo-group of Arg 22 with GC5. The result was that by only changing its side-chain torsion angles, Gln 18 can make a perfectly stable hydrogen bond to GC7. It is likely that this base pair is recognized simultaneously by Gln 18 and Ser 21. Arg 22 can also make a hydrogen bond to GC5 but it did not seem very stable and was fluctuating during the MD trajectory. This may be an imperfection of the model caused by keeping the DNA fixed in a B-conformation.

In conclusion, it can be said that a model for the *lac* repressor-operator complex has been obtained which is in agreement with experimental NOE data. The resolution is estimated to be 3.5 Å, assuming more or less rigid protein and DNA backbones. The assumption of a B-DNA type structure is justified by the overall pattern of intra-DNA NOEs, but some local deformations cannot be excluded. These would undoubtedly modify the pattern of H-bond (and hydrophobic) contacts as seen in the present simulations. The residual NOE violation is small, and falls within the uncertainty margin due to possible conformational changes of the operator.

References

1. Ernst, R.R., Bodenhausen, G. and Wokaun, A., Principles of Nuclear Magnetic Resonance in One and Two Dimensions, Clarendon Press, Oxford, 1987.
2. Wüthrich, K., NMR of Proteins and Nucleic Acids, Wiley, New York, NY, 1986.
3. Oschkinat, H., Griesinger, C., Kraulis, P.J., Sorensen, O.W., Ernst, R.R., Gronenborn, A.M. and Clore, G.M., Nature, 332 (1988) 374.
4. Vuister, G.W., Boelens, R. and Kaptein, R., J. Magn. Reson., (1988) in press.
5. Noggle, J.H. and Schirmer, R.E., The Nuclear Overhauser Effect - Chemical Applications, Academic Press, New York, NY, 1971.
6. Havel, T.F., Crippen, G.M. and Kuntz, I.D., Biopolymers, 18 (1979) 73.
7. Braun W. and Gō, N., J. Mol. Biol., 186 (1985) 611.
8. Van Gunsteren, W.F., Kaptein, R. and Zuiderweg, E.R.P., In Olson, W.K. (Ed.) Nucleic Acid Conformation and Dynamics, Report of Nato/CECAM Workshop, Orsay, 1983, pp. 79-92.
9. Kaptein, R., Zuiderweg, E.R.P., Scheek, R.M., Boelens, R. and van Gunsteren, W.F., J. Mol. Biol., 182 (1985) 179.
10. Clore, G.M., Gronenborn, A.M., Brünger, A.T. and Karplus, M., J. Mol. Biol., 186 (1985) 435.
11. Braunschweiler, L. and Ernst, R.R., J. Magn. Reson., 53 (1983) 521.
12. Davis, D.G. and Bax, A., J. Am. Chem. Soc., 107 (1985) 2821.

13. Jeener, J., Meier, B.H., Bachmann, P. and Ernst, R.R., J. Chem. Phys., 71 (1979) 4546.
14. Bothner-By, A.A., Stephens, R.L., Lee, J., Warren, C.D. and Jeanloz, R.W., J. Am. Chem. Soc., 106 (1984) 811.
15. Bax, A. and Davis, D.G., J. Magn. Reson., 63 (1985) 207.
16. Zuiderweg, E.R.P., Boelens, R. and Kaptein, R., Biopolymers, 24 (1985) 601.
17. Hyberts, S.G., Märki, W. and Wagner, G., Eur. J. Biochem., 164 (1987) 625.
18. Keepers, J.W. and James, T.L., J. Magn. Reson., 57 (1984) 404.
19. Boelens, R., Koning, T.M.G. and Kaptein, R., J. Mol. Struct., 173 (1988) 299.
20. De Vlieg, J., Boelens, R., Scheek, R.M., Kaptein, R. and van Gunsteren, W.F., Isr. J. Chem., 27 (1986) 181.
21. Neuhaus, D., Wagner, G., Vasak, M., Kägi, J.H.R. and Wüthrich, K., Eur. J. Biochem., 151 (1985) 257.
22. Blumenthal, L.M., Theory and Applications of Distance Geometry, Chelsea, New York, NY, 1970.
23. Havel, T.F., Kuntz, I.D. and Crippen, G.M., Bull. Math. Biol., 45 (1983) 665.
24. Havel, T.F. and Wüthrich, K., J. Mol. Biol., 182 (1985) 281.
25. Havel, T.F. and Wüthrich, K., Bull. Math. Biol., 45 (1984) 673.
26. Wagner, G., Braun, W., Havel, T.F., Schaumann, T., Gō, N. and Wüthrich, K., J. Mol. Biol. 196 (1987) 611.
27. De Vlieg, J., Scheek, R.M., van Gunsteren, W.F., Berendsen, H.J.C., Kaptein, R. and Thomason, J., Proteins, 3 (1988) 209.
28. Scheek, R.M. and Kaptein, R., In Oppenheimer, N.J. and James, T.L. (Eds.) NMR in Enzymology, Academic Press, New York, NY, 1988, in press.
29. Brünger, A.T., Clore, G.M., Gronenborn, A.M. and Karplus, M., Proc. Natl. Acad. Sci. U.S.A., 83 (1986) 3801.
30. Zuiderweg, E.R.P., Kaptein, R. and Wüthrich, K., Eur. J. Biochem., 137 (1983) 279.
31. Zuiderweg, E.R.P., Scheek, R.M. and Kaptein, R., Biopolymers, 24 (1985) 2257.
32. Zuiderweg, E.R.P., Kaptein, R. and Wüthrich, K., Proc. Natl. Acad. Sci. U.S.A., 80 (1983) 5837.
33. Boelens, R., Scheek, R.M., Dijkstra, K. and Kaptein, R., J. Magn. Reson., 62 (1985) 378.
34. Zuiderweg, E.R.P., Scheek, R.M., Boelens, R., van Gunsteren, W.F. and Kaptein, R., Biochimie, 67 (1985) 707.
35. Gilbert, W. and Maxam, A., Proc. Natl. Acad. Sci. U.S.A., 70 (1973) 3581.
36. Kaptein, R., Scheek, R.M., Zuiderweg, E.R.P., Boelens, R., Klappe, K.J.M., van Boom, J.H., Rüterjans, H. and Beyreuther, K., In Clementi E. and Sarma, R.H. (Eds.) Structure and Dynamics: Nucleic Acids and Proteins, Adenine Press, New York, NY, 1983, pp. 209-225.
37. Boelens, R., Scheek, R.M., van Boom, J.H. and Kaptein, R., J. Mol. Biol., 193 (1987) 213.
38. Boelens, R., Scheek, R.M., Lamerichs, R.M.J.N., de Vlieg, J., van Boom, J.H. and Kaptein, R., In Guschlbauer, W. and Saenger, W. (Eds.) DNA-Ligand Interactions, Plenum, New York, NY, 1987, pp. 191-215.
39. Barkley, M.D. and Bourgeois, S., In Miller, J.H. and Reznikoff, W.S. (Eds.) The Operon, Cold Spring Harbor Press, Cold Spring Harbor, NY, 1978, pp. 177-220.
40. Anderson, J.E., Ptashne, M.G. and Harrison, S.C., Nature, 326 (1987) 846.
41. Pabo, C. and Sauer, R., Annu. Rev. Biochem., 53 (1984) 293.
42. Weber, I.T., McKay, D.B.M. and Steitz, T.A., Nucleic Acids Res., 10 (1982) 5085.
43. Matthews, B.W., Ohlendorf, D.H., Anderson, W.F. and Takeda, Y., Proc. Natl. Acad. Sci. U.S.A., 79 (1982) 1428.

44. Ebright, R.H., J. Biomol. Struct. Dyn., 3 (1985) 281.
45. Rullmann, J.A.C., Boelens, R. and Kaptein, R., In Gralla, J. (Ed.) DNA-Protein Interactions in Transcription (UCLA Symposia Series, Vol. 95), Alan R. Liss. Inc., New York, NY, 1988, in press.
46. Billeter, M., Havel, T.F. and Kuntz, I.D., Biopolymers, 26 (1987) 777.

Subject index

Ab initio potential surface 158
Absolute temperature 4
Acceptance ratio 31
Accessibility
  solvent 175
Accuracy 96
Accuracy limiting factors 39
Accuracy of sampling 54
Actinomycin 90
Actinomycin D-DNA complex 89
Adequate sampling 53
Adiabatic charging process 134
Alanine dipeptide 96, 146, 147
Algorithm
  distance bounds driven dynamics 200
  ellipsoid 212
  variable target function 200
Alkanes
  cyclic 186
Amides 151
Amplitude 184
Analytic methods 94
Analytical continuation 9
Analytical derivative 51
Analytical form 158
Anharmonicity 161, 164
Annihilation of atoms 50
Apolar molecules 140
Approximations 39, 57, 116, 152
Assignments 207
Assumptions 57, 99
Atoms
  annihilation 50
  creation 50
Average quantity 75
Azurin 173

Backwards integration 17, 18
Barriers
  high-energy 52
B-DNA 214
Binding
  ligand/drug 135
  modes of 84
  to DNA 89, 202
Binding constants 27
  relative 39
Binding site
  copper 173
  design 173
Biological systems 73
Biomolecular structure 190, 194
64-Bit precision 110
Block average 6
Boltzmann factor 31
Boltzmann probability 4
Boltzmann's constant 4, 28
Bond-angle interaction 30
Bond-stretching interaction 30
Born formula 55
Boundary conditions 55
  extended wall region 56
  periodic 56, 116
Bragg angle 190
Butane 63, 79, 96
t-Butyl chloride 65

Calibration distance 198
Canonical
  ensemble 5, 28
  transformations 44
  variables 44
Carboxylic acids 151
Cavity formation 19
Center-of-mass motion 110
CHARMM parameter set 79
Chirality constraints 199
Cold virus 98
Compensation of errors 57
Complex 209
  actinomycin D-DNA 89
  inhibitor-enzyme 55

lac headpiece-operator 207
Complexation 65, 70
Computer-aided modeling
biomolecular structure 168
Configuration space
multi-dimensional 29
Configurational energy 4
Configurational free energy 4
Configurational integral 8, 43
Configurational partition function 3
Conformation
cyclic structure 183
distribution of 187
DNA 208
low-energy 183
shape-guided generation 183
Conformational energy surfaces 62
Conformational isomers 61
Conformational searching 103
Conjugate coordinates 40
Conjugate momenta 41
Connection formula 75, 76
Connectivity 204
Consistent force field 150
Constant pressure 55
Constant temperature 55
Constant volume 55
Constrained positions 48
Constraint forces 48
Constraints 44
Continuous change 34
Continuous coupling 52
Convergence 6, 82, 104, 107, 110, 200
radius of 191
Convergence profile 6
Continuous integration 106
Coordinates 28, 74
conjugate 40
d- 36
fractional 190
generalized 39, 40, 43
out-of-plane 164
polar 183
R- 20
reaction 64, 77
topographical transition 10
Coordinate system
choice of 42
Copper binding site 173
Corrections 110
metric tensor 47
quantum 50
Correlated spectroscopy 196
Correlation time 198
rotational 54
Coulomb interaction 30, 54
Coupling constants 199
Coupling parameter 9
λ 32, 36, 41, 48
approach 8
Covariance matrix 29
Creation of atoms 50
Criteria for sufficient sampling 54
Cross terms 162, 164
Cross-peaks 203
Cross-relaxation 194
Crystal properties 151
Crystal structure 156
Crystallin
β 169
γ 169, 179, 180
Crystallographic refinement 190
CRYSTANOVA 168, 179, 180
Cut-off distance 54, 116
Cyclic alkanes 186
Cyclic structure 183
Cyclodecane 188
Cycloheptadecane 186
Cyclononane 188
Cyclooctane 186, 188

d-coordinate 36
Database
protein structures 169
λ Dependence of the Hamiltonian 37, 52, 106
Derivative
analytic 51
numerical 51
Dielectric constant
effective 102
Dielectric relaxation time 54
Diffusive system 29
Dihedral angle constraints 202
Dihedral-angle interaction 30
Dihydrofolate reductase 73, 81
Dipoles
induced 124
Langevin 122

Dipole-dipole interaction 55
Direct methods 17
Discrete overlapping perturbation
  technique 52
Distance bounds driven dynamics 199
Distance bounds driven dynamics
  algorithm 200
Distance geometry 199
Distance matrix 199
Distribution function
  probability 29
  radial 19
Distribution of conformation 187
DNA 209
  B- 214
  conformation 208
Double-wide sampling 18
Drug design 89

Edge effects 56
Effective dielectric constant 102
Effective nonbonded interaction 30
Eglin-c 190
Electron density 153, 190
Electron density map 190
Electrostatic energy of charged groups 133
Electrostatic free energy of solvation 124
Electrostatic parameters 153
Ellipsoid algorithm 212
Empirical valence bond method 128
Energy
  criterion 31
  difference 35
  function 30, 102
  gradients 159
  kinetic 40
  minimization 103, 183
  potential 102, 128, 201
  relative solvation 61
  restraining potential 36
  second derivatives 159
  strain 156
  sublimation 153
Energy terms
  intramolecular 53
Ensemble
  average 28, 34, 104
  canonical 5, 28
  grand canonical 5
  of configurations 30
Enthalpy 5
Entropic contribution 76
Entropy 28, 34
Entropy difference 34
  integration formula 35
Environment
  relaxation time 53
Enzyme 39, 135
Enzyme catalysis 140
Errors 99, 113
  compensation of 57
Ethane 79
Excess free energy 7
Extended wall region 56

Fast Fourier refinement 190
Fitting procedure 163
Force 31
Force constant 37
Force field 30, 129, 186
  consistent 150
  intermolecular 150
Force-field parameters 56
Formate anions 160
Formula
  Born 55
  connection 75, 76
  integration 34, 38
    between two non-equilibrium
      states 36
  perturbation 34
Forwards integration 17, 18
Fractional coordinates 190
Free energy 1, 27, 60, 89
  biological systems 139
  change 75
  determination
    inaccuracy in 17
  difference 15, 33, 34, 61, 104, 140
  difference, Gibbs 55
  electrostatic 124
  excess 7
  Gibbs 4
  Helmholtz 4, 19, 28, 33
  in the absence of the restraining
    potential 37
  in the presence of the restraining
    potential 37

introduction 1
macromolecules 120, 139
of solvation 60, 120, 122, 124
perturbation 101, 126
problems 27, 94, 101
rigid model 46
surface 74, 77, 86
Free induction decay 194
Friction coefficient 32
Frictional force 32

Gas constant 39, 104
Gaussian distribution
multivariate 29
Generalized coordinates 39, 40, 43
Generalized momenta 39, 40
Generic shapes 183
Geometric constraints 199
Gibbs free energy 4
Gibbs free energy difference 55
Grand canonical ensemble 5

Half-harmonic potential 202
Hamiltonian 28, 131
λ dependence 37, 52, 106
equations of motion 74
formalism 40
Hard variables 45
Hartree-Fock calculations 159
Heat capacity 4
α-Helices 203
Helmholtz free energy 4, 19, 28, 33
Heterogeneous systems 30
High-energy barriers 52
Homonuclear Hartmann-Hahn 196
HRV-14 98
Hydration 19, 79
ion 55
Hydrocarbons 156
Hydrogen bond 150
Hydrophobic effect 63
Hydrophobic interaction 23
Hysteresis 53, 64, 113

Importance sampling 21, 61
Improper torsion 165
Indices of the reflection 190
Induced dipoles 124
Inhibitor 39
Inhibitor-enzyme complex 55
Insufficient sampling 107
Integration
backwards 17, 18
formula, between two non-equilibrium states 36
forwards 17, 18
window of 23
Integration formula 34, 36,38
Integration methods 32
Integration time-step 112
Interaction
bond-angle 30
bond-stretching 30
Coulomb 54
dihedral-angle 30
dipole-dipole 55
hydrophobic 23
long-range, treatment of 54
potential 30
protein-ligand 140
torsional 30
van der Waals 30
Intermediate state 34
Intermolecular force field 150
Ions 141
Ion hydration 55
Ion pair 65, 122
solvent separated 66
Isomers
conformational 61

Jacobian transformation 21, 45

Kinetic energy 40

lac headpiece-operator complex 207
lac operon 207
lac repressor headpiece 195
Langevin dipoles 122
Langevin equation of motion 32
Linear behavior 53
Loop selection 173
Low-energy conformation 183

Machine
precision 107
special-purpose 143
Major groove 211
Mass-metric tensor 43, 45
Mean energy 4

Mean force
  potential of 9, 19, 20, 23, 61, 63, 86
Methane 70, 115
  dimer 70
Methods
  analytic 94
  direct 17
  empirical valence bond 128
  Monte Carlo 1, 31
  nonlinear 15
  particle insertion 29
  perturbation 16
  probability ratio 23
  probe 29
  subinterval perturbation 18
  systematic search 31
  thermodynamic cycle-perturbation 91
  thermodynamic integration 32
  thermodynamic perturbation 32
Methotrexate 142
N-Methylacetamide 64
Methylammonium acetate 68
Metric matrix distance geometry 199
Metric tensor corrections 47
Minimum energy reaction path 60
Model building 212
Molecular dynamics 31, 77, 103
  refinement 190, 191
  restrained 201
Molecular potential surface 161
Molecular simulation 1, 3
Momentum 28, 74
  generalized 39, 40
Monte Carlo method 1, 31
Morse potential 130
Multi-dimensional configuration space 29
Multiple minima problem 25
Multivariate Gaussian distribution 29
Mutations 175

N-particle system 4
NMR
  data 90
  spectroscopy 194
  three-dimensional 194
  two-dimensional 194
Newton's equation of motion 31, 74, 103, 190, 201
Noble gases 113, 141
Node number 184
Non-Boltzmann sampling 21
Non-bond mixing rule 110
Non-chemical processes 39
Non-conjugate velocity 42
Non-constrained atomic positions 48
Non-diffusive system 29
Non-equilibrium state 36, 38
Nonlinear methods 15
Non-NOE information 199
Normal mode analysis 149
Nuclear Overhauser effect 194
  data 90
Nucleophilic substitution reaction 23
Number density 19
Numerical
  derivative 51
  precision 107
  quadrature 75

OPLS parameter set 64, 79
Out-of-plane coordinate 164

Parameter set
  CHARMM 79
  OPLS 64, 79
Parameters
  electrostatic 153
  force field 56
  potential function 62
  thermal 190
Parameterization 53
Particle insertion method 29
Partition function 28, 33
  configurational 3
Pattern-recognition procedure 210
Periodic boundary conditions 56, 116
Perturbation formula 34
Perturbation method 15, 16, 34
  discrete overlapping 52
  subinterval 18
  thermodynamic cycle 73, 91
Phase 184
Phase space probability 28
pKa 124
Planck's constant 28
Plastocyanin 173
Poisson equation 126
Polar coordinates 183
Polarizability 57
Polarization 62

Polygon 184
Potential energy 102, 128, 201
Potential energy functions
  biomolecular systems 149
  organic systems 149
Potential functions 62, 150
  restraining 38
Potential function parameters 62
Potential of mean force 9, 19, 20, 23, 61, 63, 86
Precision
  64-bit 110
  machine 107
Preferential sampling 63
Pressure 5
  constant 55
Primitive model 66
Probability distribution
  direct determination 29
  function 29
λ Probability function 12
Probability ratio method 23
Probe method 29
Processes
  non-chemical 39
Propane 79
Protein design
  CRYSTANOVA 168
  knowledge-based 168
Protein-ligand interaction 140
Proton-proton distances 194
Pseudoatoms 198
Pseudopotentials 202

Quantum corrections 50
Quantum mechanical calculations 135
Quartic terms 161
Quasi-ergodic problem 14, 17
Quasiharmonic approximation 2

R-coordinate 20
R-factor 192
Radial distribution function 19
Radiofrequency 196
Radius of convergence 191
Random displacement 31
Range of sampling 21
Range of validity 99
Reaction
  nucleophilic substitution 23
Reaction coordinate 64, 77
Reaction path
  minimum energy 60
Reaction profile 131
Reactive trajectory 132
Recognition helix 211
Reference positions 48
Refinement
  crystallographic 190
  fast Fourier 190
  molecular dynamics 190, 191
  restrained 191
  structure 201
Relative binding constant 39
Relative solvation energy 61
Relaxation matrix treatment 198
Relaxation time 54
Relaxation time of the environment 53
Reliability 57, 102
Repressor protein 202
Repressor-operator complex 207
Resonance assignment 195
Restrained molecular dynamics 201
Restrained refinement 191
Reversibility 53
Reversible process 145
Reversible work 19, 55
Rigid model 45
  free energy 46
  partition function 46
Ring shape 184
Ring system 183
Rotamers 185
Rotational correlation time 54

Salt bridge 68
Sampling 105
  accuracy of 54
  adequate 53
  criteria 54
  double-wide 18
  importance 21, 61
  insufficient 107
  non-Boltzmann 21
  preferential 63
  range of 21
  umbrella 21, 22, 63
  window 23
Sampling properties 205
Scattering factor 190

Second derivatives 157
Sequence 176
Sequence similarity 180
Sequential assignment 196
SHAKE 47
Shapes
  generic 183
Side-chain conformations 169
Slow growth 13, 15, 17, 52, 98, 106
Soft variables 45
Solvation free energy 122
Solvent accessibility 175
Solvent effect 97
Solvent-separated ion pair 66
Special-purpose machine 143
Spectrum 210
Spin-diffusion 198
Spin-relaxation phenomenon 194
S-Protein 179, 180
Statistical
  significance 79
  uncertainty 6
    problem 7
Stereospecific assignments 198
Stochastic dynamics 31
Strain energy 156
Structure
  biomolecular 190, 194
  from NMR 194
  from X-ray 190
Structure factor amplitude 190
Structure factor equation 190
Structure refinement 201
Subinterval 18
Subinterval perturbation method 18
Sublimation energy 153
Subtilisin 191
Superoxide dismutase 173
Surface constraints 126
Systematic search method 31
Systems
  biological 73
  diffusive 29
  heterogeneous 30
  non-diffusive 29

Technique
  continuous coupling 52
  slow growth 13, 15, 17, 52, 98, 106
  window 52, 61, 75, 106
Temperature 28, 104
  constant 55
  derivative 77, 85
  factor 190
Tensor
  mass-metric 43, 45
Thermal bath 202
Thermal parameters 190
Thermitase 190, 191
Thermodynamic cycle perturbation
  method 73, 91
Thermodynamic cycle technique 39
Thermodynamic
  cycles 38
  derivative property 74
  integration 9, 12, 13, 32, 52, 73
  integration formula 51
  pathway 74
  perturbation 32, 52
  simulation 73
Three-dimensional NMR 194
Time
  dielectric relaxation 54
  relaxation 54
  rotational correlation 54
Time-reversal invariant formula 52
Topographical transition coordinates 10
Torsion
  improper 165
Torsional interaction 30
Total correlation spectroscopy 196
Trans-gauche equilibrium 63
Treatment of long-range interactions 54
Trimethoprim 81
Two-dimensional NMR 194

Umbrella sampling 21, 22, 63
Uniqueness 205

Valence bond 127
Van der Waals interaction 30
Variable
  canonical 44
  hard 45
  soft 45
Variable target function algorithm 200
Vibrational spectra 156
Vicinal J-coupling 196
Violation
  of NOE distances 204

Volume
  constant 55

Water
  TIP4P model 64
Weight factor
  λ dependent 47
Window of integration 23
Window sampling 23
Window technique 52, 61, 75, 106
Work
  reversible 19, 55

X-ray structure 190